Laboratory Experiments

HOLT

Physics

$$\sin\theta_c = \frac{n_r}{n_i}$$

PHYSICS INTERACTIVE TUTOR

SERWAY · FAUGHN

HOLT, RINEHART AND WINSTON

Harcourt Brace & Company

Austin · New York · Orlando · Atlanta · San Francisco · Boston · Dallas · Toronto · London

Holt Physics

Laboratory Experiments

Lab Authors

Douglas W. Biedenweg, Ph.D.
Chadwick School
Palos Verdes, CA

Kaye M. Elsner-McCall
Physics Teacher
Riverwood High School
Fulton County Schools
Atlanta, GA

Anthony L. Komon
Physics Teacher
Science Department
Niskayuna High School
Schenectady, NY

Sean P. Lally
Chairman of Science
Sewickley Academy
Sewickley, PA

Safety Reviewer

Gregory Puskar
Laboratory Manager
Physics Department
West Virginia University
Morgantown, WV

Laboratory Reviewers

Lee Sennholtz
Central Scientific Company
Franklin Park, IL

Martin Taylor
Central Scientific Company
Franklin Park, IL

Printed in the United States of America

ISBN 0-03-051864-4

6 095 02 01 00

Contents

Using the labs in this book

Taking different approaches to the challenge of physics

The *Holt Physics Laboratory Experiments* booklet contains 33 all-new laboratory experiments. The two types of labs in this booklet are designed to help you learn physics from the beginning of each chapter to the end. You will probably find that the labs in this booklet are organized differently from those in the textbook and from any laboratory experiments you have done before. The first type of lab, called a *Discovery Lab*, guides you through new lessons with a step-by-step, hands-on approach that gives you real-world experience with the physics concepts you will study in each chapter. The second type of lab is called an *Invention Lab*, and it gives you the opportunity to use your physics knowledge by developing an invention or process to solve a real problem.

As you work on both of these types of labs, you will develop a solid understanding of how the concepts presented in the textbook relate to everyday physical phenomena, and you will use your understanding of physics to solve problems like those faced by physicists and engineers every day.

Discovery Labs

The Discovery Labs are divided up into small sections, each presenting a basic physics concept. Each section provides step-by-step procedures for you to follow, encouraging you to make careful observations and interpretations as you perform each step of the lab. After each section, there is a series of questions designed to help you make sense of your observations and data and relate them to the physics concepts you will study in the chapter.

What you should do before a Discovery Lab

Preparation will help you work safely and efficiently. Before a lab, be sure to do the following:

- **Read the lab procedure** to make sure you understand what you will do in each step.

- **Read the safety information** that begins on page ix, as well as the special safety instructions provided in the lab procedure. Plan to wear appropriate shoes, clothing, and protective safety equipment while you work in the lab.

- **Write down any questions** you have in your lab notebook and ask them before the lab begins.

- **Prepare all necessary data tables** so that you will be able to concentrate on your work when you are in the lab.

What you should do after a Discovery Lab

Most teachers require a written lab report as a way of making sure that you understood what you were doing in the lab. Your teacher will give you specific details about how to organize your written work for the Discovery Labs, but most lab reports will include the following:

- **the title** of the lab

- **data tables and observations** that are organized, complete, and easy to understand

- **answers** to the items and questions that appear after each section of the procedure

Invention Labs

The Invention Labs may seem unusual to you because they do not provide you with step-by-step instructions. The Invention Labs present problems in the context of assignments for an engineering and research company. These labs refer to you as an employee of the company, and your teacher has the role of a supervisor. Lab situations are given for real-life problems. The Invention Labs require you to develop your own procedure to solve a problem presented to your company by a client. As part of the research and development team working for the client, you must choose equipment and a procedure to solve the problem. Each lab is designed to use physics concepts that you have studied in the previous chapters, and each lab contains hints and useful information about how to solve the problem.

What you should do before an Invention Lab

Before you will be allowed to work on the lab, you must turn in an initial plan. Your teacher will tell you exactly how to write an initial plan, but most plans must include a detailed description of the procedure you plan to use, the measurements and observations you will take, and a list of equipment you will use to complete the lab. Your teacher, acting as your supervisor, must approve your plan before you are allowed to proceed. Before you begin writing an initial plan, complete the following steps:

- **Read the Invention Lab thoroughly** to make sure you understand the problem. Read carefully, and pay attention to the hints and guidelines that are presented in the lab.

- **Jot down notes** in your lab notebook as you find clues and begin to develop a plan.

- **Consider how to use physics concepts** to solve the problem. Think about the measurements and observations you will have to make to find a solution.

- **Imagine working through a procedure,** keeping track of each step and the equipment you will need. Pay special attention to safety issues.

- **Carefully consider** ways to improve your approach, in terms of logic, safety, and efficiency.

- **Read the safety information** that begins on page ix, as well as the special safety instructions provided in the lab. Plan to wear appropriate shoes, clothing, and protective safety equipment while working in the lab.

What you should do after an Invention Lab

When you have completed the lab, you will present your results in the form of a Patent Application. Your teacher may have additional requirements for your report. A sample Patent Application lab report can be found on page vii.

The format for the Patent Application lab report is based on the real requirements for patent applications in the United States. For the Invention Lab reports, a Patent Application must include the following eight sections.

1. **Date, Title, and Inventor:** The date and title of the invention and the name of the principal inventor, followed by the names of any team members or joint inventors. If your team is preparing a single application, all members may be listed jointly.

2. **Background—Field of Invention:** A sentence that states both the general and specific field relating to your invention. For example, "This invention relates to direct current circuits, specifically to decorative lighting."

3. **Drawings:** Include as many types of drawings from as many perspectives as you need to present the mechanics of your invention. Each part of your invention should be labeled with a number or letter in the drawing for easy reference.

4. **Description of Drawings:** A brief description of each drawing, specifying the type of view being presented (cross-section, top view, side view, schematic, exploded, etc.).

5. **List of Reference Numerals:** This is a list of the numbers you used in your drawings, with a description of what each number refers to.

6. **Description of Invention:** This is a detailed description of all the parts of the invention. Refer to your diagrams. Describe the individual parts and how they are connected.

7. **Operation of Invention:** Describe the actual operation of your invention. Include a discussion of the theory of how it operates. Include any equations, proportions, or formulas necessary for an understanding of how your invention works. Also include the physical values you measured in the lab. (Hint: It is always helpful to proceed with the description in an orderly fashion—for example, when describing an electrical circuit, you may want to begin at the negative post of the battery and "follow" the current through the circuit.)

8. **Conclusion, Ramifications, and Scope of Invention:** One sentence restates the purpose and operation of the invention. The rest of this section is a discussion of possible variations of the design and can include ideas for other possible applications of the device or process.

Sample Patent Application Lab Report

This sample lab report is provided to give you a model to follow. Your patent applications will not be exactly like this one, but they should contain the same basic parts, as described above.

1. **Date:** May 18, 1999
   Title: Doormat Lighting System
   Inventor: Antonia Briggs
                  Sinh Ngyuen

2. **Background—Field of Invention:** This invention relates to resistors in direct current circuits, specifically to security lighting.

3. **Drawings:**

Drawing A

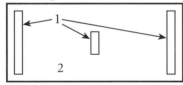

Drawing B

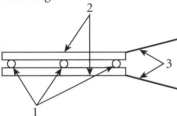

Drawing C

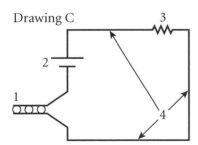

4. **Description of Drawings** and

5. **List of Reference Numerals:**

   Drawing A   Top view of the bottom of the doormat

   1. plastic drinking straws, two on ends are 8 cm long, one in center is 4 cm long

   2. flat pieces of cardboard, 20 cm × 10 cm, covered with heavy-duty aluminum foil

   Drawing B   Side view of the doormat

   1. side view of three plastic drinking straws (see Drawing A: 1)

   2. side view of two pieces of cardboard, 20 cm × 10 cm, both covered with aluminum foil

   3. connecting wires, connected to top side of aluminum foil

   Drawing C   Circuit diagram

   1. Doormat (see Drawing B: 1, 2, and 3)
   2. dc battery
   3. light bulb
   4. insulated connecting wires

6. **Description of Invention:** The Doormat Lighting System consists of a doormat wired in a series circuit with a dc battery and a light bulb. The doormat is constructed by taking two pieces of cardboard or other firm material and covering them with aluminum foil. The aluminum foil is glued or taped securely to the cardboard. On one piece of cardboard, drinking straws are glued securely at each end and in the middle of the cardboard. The straws on the ends should be almost the same length as the end of the cardboard, and the straw in the middle should be about half that length. All straws should be centered lengthwise on the cardboard, so there is equal distance from the end of the straw to the edge of the cardboard on both sides. See Drawing A.

   The second foil-covered piece of cardboard is placed on top of the straws and glued securely. The cardboard pieces should be stacked so that the edges line up exactly, and the straws should prevent them from touching. Insulated connecting wires are attached to the top and bottom of the foil-covered pieces. See Drawing B for a side view.

   The insulated connecting wires are used to wire the doormat in series with a dc battery and a light bulb, as shown in Drawing C.

7. **Operation of Invention:** The Doormat Lighting System will light a lightbulb when weight is applied to the doormat. The purpose of this invention is to allow a person to step on the doormat and turn on the light. The dc battery connected in

series with the light bulb and the doormat provides a potential difference to the circuit. The doormat acts as a switch in this circuit. When the two surfaces of the doormat are not touching, as when no weight is applied to the doormat, the switch is open and there is no current in the circuit. When weight is applied to the doormat and the two foil-covered surfaces touch, the switch is closed and there is current in the circuit according to the potential difference and the resistance present in the circuit. This relationship is given by the following equation:

$$I = \frac{\Delta V}{R}$$

We developed our circuit using a 9 V dc battery and a 6.3 V, 115 mA light bulb.

When there is current in the circuit, the light bulb, which acts as a resistor in the circuit, will light.

When the weight is removed from the doormat, the plates of the doormat will separate, opening the switch, and there will be no current in the circuit. The light bulb will no longer be lighted.

8. **Conclusion, Ramifications, and Scope of Invention:** The Doormat Lighting System is a security lighting device that uses a resistor in a dc circuit with a battery. The doormat itself operates as a switch in this circuit, and a light bulb operates as a resistor. When the doormat is stepped on, the switch is closed and the light bulb lights. When the weight is removed from the doormat, the switch is opened and the light stops.

In this design, pieces of plastic drinking straws are used to separate the two conducting parts of the doormat. Other items, such as springs, may be used in place of the drinking straws. Any material used for this purpose must be flexible, so that it will compress when weight is applied and will return to its original position when the weight is removed, and it must not conduct electricity. In fact, another type of separator may be better, because the drinking straws become flattened with use and will need to be replaced often to maintain the required distance between the two pieces.

The dimensions of all the parts of this system, from the size of the doormat to the length of the wires, depends upon the desired use. This system may be used to place a doormat outside the door of a house and light a lamp above the door, or it may be used to light a lamp placed inside the house or at another location. The battery and light bulb must be selected so that the battery provides enough potential difference to light the selected bulb but not enough to cause a fire or short circuit.

Another possible use of the design would be to use a resistor other than a light bulb. For example, the circuit could contain a buzzer or some other device. In this way, the circuit could operate as an alarm system or a doorbell. In addition, the switch in the circuit could be designed for use in any device that required a pressure-sensitive switch. The switch could be placed in the bottom of a mailbox and wired to a light or buzzer inside the house; this system would notify someone inside the house that the mail had been delivered.

Because aluminum foil conducts electricity, it would be necessary to cover the entire switch in insulating material before using this device.

Safety in the Physics Laboratory

Lab work is the key to progress in science. Therefore, systematic, careful lab work is an essential part of any science program. In this class, you will practice some of the same fundamental laboratory procedures and techniques that experimental physicists use to pursue new knowledge.

The equipment and apparatus you will use involve various safety hazards, just as they do for working physicists. You must be aware of these hazards. Your teacher will guide you in properly using the equipment and carrying out the experiments, but you must also take responsibility for your part in this process. With the active involvement of you and your teacher, these risks can be minimized so that working in the physics laboratory can be a safe, enjoyable process of discovery.

These safety rules always apply in the lab

1. **Always wear a lab apron and safety goggles.**
    Wear these safety devices whenever you are in the lab, not just when you are working on an experiment.

2. **No contact lenses in the lab.**
    Contact lenses should not be worn during any investigations using chemicals (even if you are wearing goggles). In the event of an accident, chemicals can get behind contact lenses and cause serious damage before the lenses can be removed. If your doctor requires that you wear contact lenses instead of glasses, you should wear eye-cup safety goggles in the lab. Ask your doctor or your teacher how to use this important eye protection.

3. **Personal apparel should be appropriate for laboratory work.**
    On lab days avoid wearing long necklaces, dangling bracelets, bulky jewelry, and bulky or loose-fitting clothing. Long hair should be tied back. Loose, dangling items may get caught in moving parts, accidentally contact electrical connections, or interfere with the investigation in a potentially hazardous manner. In addition, chemical fumes may react with some jewelry, such as pearls, and

ruin them. Cotton clothing is preferable to wool, nylon, or polyester. Wear shoes that will protect your feet from chemical spills and falling objects—open-toed shoes or sandals, and shoes with woven leather straps are not allowed in the laboratory.

4. **NEVER work alone in the laboratory.**
    Work in the lab only while under the supervision of your teacher. Do not leave equipment unattended while it is in operation.

5. **Only books and notebooks needed for the experiment should be in the lab.**
    Only the lab notebook and the textbook should be used. Keep other books, backpacks, purses, and similar items in your desk, locker, or designated storage area.

6. **Read the entire experiment before entering the lab.**
    Your teacher will review applicable safety precautions before the lab. If you are not sure of something, ask your teacher about it.

7. **Always heed safety symbols and cautions written in the experimental investigations and handouts, posted in the room, and given verbally by your teacher.**
    They are provided for your safety.

8. **Know the proper fire drill procedures and the location of fire exits and emergency equipment.**
    Make sure you know the procedures to follow in case of a fire or an emergency.

9. **If your clothing catches on fire, do not run; WALK to the safety shower, stand under it, and turn it on.**
    Call to your teacher while you do this.

10. **Report all accidents to the teacher immediately, no matter how minor.**
    In addition, if you get a headache, feel sick to your stomach, or feel dizzy, tell your teacher immediately.

11. **Report all spills to your teacher immediately.**
    Call your teacher rather than trying to clean up a spill yourself. Your teacher will tell you if it is safe for you to clean up the spill; if not, your teacher will know how the spill should be cleaned up safely.

12. **Student-designed inquiry investigations, such as the Invention Labs in the Laboratory Experiments manual, must be approved by the teacher before being attempted by the student.**

13. **DO NOT perform unauthorized experiments or use materials and equipment in a manner for which they were not intended.**

    Use only materials and equipment listed in the activity equipment list or authorized by your teacher. Steps in a procedure should only be performed as described in the textbook or lab manual or approved by your teacher.

14. **Stay alert in the lab, and proceed with caution.**

    Be aware of others near you or your equipment when you are performing an experiment. If you are not sure of how to proceed, ask.

15. **Horseplay in the lab is very dangerous.**

    Laboratory equipment and apparatus are not toys; never play in the lab or use lab time or equipment for anything other than their intended purpose.

16. **Food, beverages, chewing gum, and tobacco products are NEVER permitted in the laboratory.**

17. **NEVER taste chemicals. Do not touch chemicals or allow them to contact areas of bare skin.**

18. **Use extreme CAUTION when working with hot plates or other heating devices.**

    Keep your head, hands, hair, and clothing away from the flame or heating area, and turn heating devices off when they are not in use. Remember that metal surfaces connected to the heated area will become hot by conduction. Gas burners should be lit only with a spark lighter. Make sure all heating devices and gas valves are turned off before leaving the laboratory. Never leave a hot plate or other heating device unattended when it is in use. Remember that many metal, ceramic, and glass items do not always look hot when they are hot. Allow all items to cool before storing.

19. **Exercise caution when working with electrical equipment.**

    Do not use electrical equipment with frayed or twisted wires. Be sure your hands are dry before using electrical equipment. Do not let electrical cords dangle from work stations; dangling cords can cause electrical shocks and other injuries.

20. **Keep work areas and apparatus clean and neat.** Always clean up any clutter made during lab work, rearrange apparatus in an orderly manner, and report any damaged or missing items.

21. **Always thoroughly wash your hands with soap and water at the conclusion of each investigation.**

Safety Symbols

The following safety symbols will appear in the laboratory experiments to emphasize additional important areas of caution. Learn what they represent so you can take the appropriate precautions. Remember that the safety symbols represent hazards that apply to a specific activity, but the numbered rules given on the previous pages apply to all laboratory work.

 Waste Disposal

- Never put broken glass or ceramics in a regular waste container. Use a dustpan, a brush, and heavy gloves to carefully pick up broken pieces, and dispose of them in a container specifically provided for this purpose.

- Dispose of chemicals as instructed by your teacher. Never pour hazardous chemicals into a regular waste container. Never pour radioactive materials down the drain.

 Heating Safety

- When using a burner or hot plate, always wear goggles and an apron to protect your eyes and clothing. Tie back long hair, secure loose clothing and remove loose jewelry.

- Never leave a hot plate unattended while it is turned on.

- Wire coils may heat up rapidly during this experiment. If heating occurs, open the switch immediately and handle the equipment with a hot mitt.

- Allow all equipment to cool before storing it.
- If your clothing catches on fire, walk to the emergency lab shower and use the shower to put out the fire.

 Hand Safety

- Perform this experiment in a clear area. Attach masses securely. Falling, dropped, or swinging objects can cause serious injury.
- Use a hot mitt to handle resistors, light sources, and other equipment that may be hot. Allow all equipment to cool before storing it.

 Glassware Safety

- If a thermometer breaks, notify the teacher **immediately.**
- Do not heat glassware that is broken, chipped, or cracked. Use tongs or a hot mitt to handle heated glassware and other equipment that may be hot. Allow all equipment to cool before storing it.
- If a bulb breaks, notify your teacher immediately. Do not remove broken bulbs from sockets.

 Electrical Safety

- Never close a circuit until it has been approved by your teacher. Never rewire or adjust any element of a closed circuit.
- Never work with electricity near water. Be sure the floor and all work surfaces are dry.
- If the pointer on any kind of meter moves off scale, open the circuit immediately by opening the switch.

- Do not work with any batteries, electrical devices, or magnets other than those provided by your teacher.

 Chemical Safety

- Do not eat or drink anything in the laboratory. Never taste chemicals or touch them with your bare hands.
- Do not allow radioactive materials to come into contact with your skin, hair, clothing, or personal belongings. Although the materials used in this lab are not hazardous when used properly, radioactive materials can cause serious illness.

 Clothing Protection

- Tie back long hair, secure loose clothing, and remove loose jewelry to prevent their getting caught in moving or rotating parts or coming into contact with hazardous chemicals.

 Eye Protection

- Wear eye protection, and perform this experiment in a clear area. Swinging objects can cause serious injury.
- Avoid looking directly at a light source. Looking directly at a light source may cause permanent eye damage.

The Circumference-Diameter Ratio of a Circle

MATERIALS
✔ cord
✔ masking tape
✔ metric rulers
✔ pencil
✔ several cylindrical objects of varying size
✔ white paper

SAFETY

- Review lab safety guidelines. Always follow correct procedures in the lab.

OBJECTIVES

- Develop techniques for measuring the circumference and diameter of a cylinder.
- Use data to construct a graph.
- Determine the slope of a graph.
- Analyze error in an experiment.

Measurements of a cylinder

Procedure

1. Select one of the cylinders. Examine the cylinder to determine how many different measurements would be necessary to give a complete description of the cylinder. In this lab, you will use a cylinder's measurements to identify one cylinder from a group of cylinders, so make sure your measurements enable you to distinguish the cylinder from similar cylinders.

2. Determine at least two different methods of making the measurements. Be sure you include ways to measure the circumference of the cylinder in each method. Keep in mind that you must measure each quantity directly; no values can be found through calculations.

3. Take all the measurements for the cylinder using the first method you developed. Record all measurements in your notebook using the appropriate SI units. Make sure to include all measured digits plus one estimated digit.

4. Place the cylinder into a container with a group of other cylinders. Trade measurements with another group. Use your method of measurement to find the cylinder that matches the measurements you were given.

Analysis

A. What measurements did you make?

B. What was your method of measuring the cylinder? Describe your method in detail.

C. Did you find the cylinder that matched the measurements you were given? If not, why not?

D. Did the other group correctly identify the cylinder you measured? If not, why not?

E. Compare your measurements with the other group's measurements for the same cylinder. Are the measurements the same? Explain any differences in your methods or measurements.

Comparing methods of measurement

Procedure

5. Using the same method you used to measure the first cylinder, measure the length, diameter, and circumference of several more cylinders. Label each cylinder with an identifying name written on masking tape. Record your measurements in your notebook using the appropriate SI units.

6. Perform another trial, using a different method to take the measurements. Repeat the measurements for the length, diameter, and circumference of all cylinders. Record your measurements in your notebook using the appropriate SI units.

Analysis

F. Compare the results you obtained using two different methods of measurement. Did you get the same measurements for each cylinder regardless of which method you used? If not, explain what you think caused the difference.

G. Which method do you think was best for measuring the cylinders? What were some problems with the other methods you tried?

H. How could you determine which method of measuring the cylinders gave the best results?

Data analysis

Procedure

7. Use the data you collected to decide which method of measuring the cylinders gave the best results. For each cylinder, select the measurements taken with this method.

8. Use the data you selected in step 7. For each cylinder, find the value for the circumference of the cylinder divided by the diameter of the cylinder.

Analysis

I. Is the relationship between the circumference and the diameter the same for all cylinders, or is it different for each one?

J. Based on your results, what measurements do you think are necessary to give a complete description of a cylinder?

Graphing data

Procedure

9. Using the data you selected, make a graph of the circumference of the cylinders versus the diameter of the cylinders. For each cylinder, plot a point on the graph that represents the cylinder's circumference and diameter.

10. Draw the line or curve that best fits the points on the graph. Not all the points on the graph will actually fall directly on the line, but the line should follow the shape made by most of the points. The line should not connect each point directly to the next one. Instead, it should be drawn as a smooth line or curve connecting most of the points.

11. Select two points on the line, one at the beginning and one at the end. Make sure the points selected are points on the best fit line but are not data points. Use the scales on the axes of the graph to determine the circumference and diameter of the cylinders that would be represented by these points on the line.

12. Label the points that you selected A and B. Find the difference between the values for the circumference at these points, and use this as the *rise*. In other words, subtract the value for the circumference at A from the value for the circumference at B. Find the difference between the values for the diameter at these points, and use this as the *run*. Subtract the value for the diameter at A from the value for the diameter at B.

13. Find the slope of the line, using the equation $slope = \frac{rise}{run}$.

Analysis

K. On your graph, which quantity is the independent variable?

L. On your graph, which quantity is the dependent variable?

M. Describe the shape of the curve in your graph.

N. What is the value that you calculated for the slope of the curve in your graph? Compare this to the relationship between the circumference and the diameter that you calculated in step 8.

O. Based on your data and your graph, do you think it is better to find the relationship between the circumference and the diameter by using the slope of the graph or by calculating individual values? Explain your answer.

Bubble Solutions

TANTRUM TOYS, INC.
TROY, NEW YORK

August 15, 1999

Ms. Elaine Taylor
Product Development Department
1% Inspiration Laboratories
14557 West Post Road
Tempe, Arizona 85289

Dear Ms. Taylor:

At Tantrum Toys, we always try to stay one step ahead of the market. That's why we are looking into new formulations for our famous bubble solution. We have developed a new formula that we believe will help our bubble solution make bigger, longer-lasting bubbles.

We would like you to test our new bubble solution against several other commercially available solutions, including the solution currently marketed by Tantrum Toys. In order to cut down on human error or bias in the laboratory, we are sending the solutions to you in identical packaging, marked only with a letter. We would like you to test all solutions to find out which produces the biggest bubbles.

Please perform two tests: the dome test and the free-floating bubble test. For the dome test, use a straw to blow a domed bubble in a pan of solution. Measure the height and diameter of each dome. For the free-floating bubble test, construct a bubble maker to make large free-floating bubbles. Measure the diameter of each bubble.

When you have finished your tests, put together a report describing how you performed the tests, showing the equipment you used, and detailing your results. Please have the report and all unused solutions delivered to my office by September 8.

Good luck,

Stewart Clydesdale

Stewart Clydesdale

A description of a bubble maker is on page 6.

1% Inspiration Laboratories

MEMORANDUM

Date: August 19, 1999
To: Product Testing Team
From: Elaine Taylor

I think the best way to get all these tests done in time is to have several people work on them at once. Hopefully, one solution will be obviously better than the others and all our results will be the same. Before you go into the lab, prepare a plan for each of the tests described in the letter. Be sure to include your plan for measuring the width, height, and diameter of the bubbles in the lab. This will be a tricky procedure, because we have to find a way to get good measurements without actually touching the bubbles. Consistency and accuracy will also be very important, especially since we will have to work quickly and carefully to make our measurements before the bubbles pop.

Present your plan to me for approval before you start work in the lab. For each test, your plan should include a list of materials needed, a diagram, and a one- or two-sentence explanation of the procedure you will use. I have included a list of the equipment we have available. If you need something that you can't find on the list, be sure to ask about it; there may be more equipment available.

For the second test, you will need to construct a bubble maker using the materials on the list. The background information Mr. Clydesdale sent me on one type of bubble wand is attached to this memo, but I will be interested to see what you can come up with. Be sure to include your design when you submit your plan for approval.

When you have all your results, write a report using the format of a patent application. Remember to document all your testing and development procedures in your lab notebook.

14557 West Post Road • Tempe, Arizona 852

See next page for safety requirements, materials list, and more hints.

continued

MATERIALS

ITEM	QTY.
✔ adhesive tape	1 roll
✔ aluminum pans	
✔ bubble solutions	
✔ cord	100 cm
✔ meterstick	1
✔ metric ruler	1
✔ paper towels	
✔ plastic drinking straws	6
✔ rubber bands	4

SAFETY

- Do not eat or drink anything in the laboratory. Never taste chemicals or touch them with your bare hands.

- Dispose of chemicals as instructed by your teacher. Never pour hazardous chemicals into a regular waste container.

- Tie back long hair, secure loose clothing, and remove loose jewelry to prevent their coming into contact with hazardous chemicals.

- Wear eye protection. Keep chemicals away from eyes.

When it comes to bubbles, the bigger the better

It may not seem like a museum piece to some people, but to the children who visit The Discovery Science Museum in Birmingham, a simple contraption made of plastic drinking straws and string is among the best things the museum has to offer.

This device allows students to make soap bubbles bigger than any they've seen before. This magic wand was invented right here at the Discovery Science Museum, but it can be recreated by children everywhere because the materials are readily available.

All you need is a piece of thread about 1 meter long and two plastic drinking straws. Thread the string through both straws, and tie the two ends of the string into a knot. Pull the string around until the knot is safely hidden away inside one of the straws. Use both hands to pull the straws apart, so that they are parallel to each other, with the strings relaxed between them. Dip the two string sides into bubble solution—either a commercial brand, like the favorite from Tantrum Toys, or a solution made with ordinary dish soap.

To make a long bubble, pull the frame through the air or blow gently. This activity will delight children immediately, but we bet it won't take adults long to admit that it is a great work of art!

Motion

MATERIALS

✔ battery-operated toy car
✔ block, book, or clay
✔ graph paper
✔ masking tape
✔ metal ball
✔ meterstick
✔ stopwatch
✔ track
✔ wooden block

SAFETY

- Tie back long hair, secure loose clothing, and remove loose jewelry to prevent their being caught in moving or rotating parts.

- Perform this experiment in a clear area. Moving masses can cause serious injury.

OBJECTIVES

- Observe objects moving at a constant speed and objects moving with changing speed.
- Graph the relationships between distance and time for moving objects.
- Interpret graphs relating distance and time for moving objects.

Moving at a constant speed

Procedure

1. Find a clear, flat surface a few meters long to perform your experiment. Make sure the area is free of obstacles and traffic. Choose a starting point for your car. Mark this point with masking tape, and label it "starting point."

2. Start the car, and place it on the starting point. Release the car (your lab partner should start the stopwatch at the same time). Let the car move in a straight line for 2.0 s. Notice where the car is after 2.0 s. Repeat for several trials, until you find the point that the car consistently crosses after 2.0 s. Mark this point with masking tape, and label it "0.00 m." Throughout this lab, you will start the car at the original starting point, but you will begin to measure the distance and time of the car's motion when the car crosses the 0.00 m mark.

3. Start the car, and place it on the floor at the starting point. Observe the car as it moves. Be sure to start the stopwatch as the car crosses the 0.00 m mark.

Start

Stop

Car

Starting Point 0.00 m

10.0 s

⊢---------- Distance Traveled ----------⊣

4. After 10.0 s, mark the position of the car with the masking tape. Label this mark "10.0 s."

5. Repeat steps 3 and 4 for 9.0 s, 8.0 s, 7.0 s, 6.0 s, 5.0 s, 4.0 s, 3.0 s, and 2.0 s. Be sure to label each point according to how much time it took for the car to get to that point from the 0.00 m mark.

6. Use the meterstick to measure the exact distance from the 0.00 m mark to each timed position mark. (Do not measure the distance from the starting point.)

7. For each position marked with tape, record the position and time in your notebook, using the appropriate SI units. Make sure to record all measured digits plus one estimated digit.

8. If your car has a multiple speed switch, set the car at a new speed and repeat steps 3–7.

Analysis

A. Did the car speed up or slow down as it traveled, or did it maintain the same speed? How can you tell?

B. Make a graph of your data with time on the *x*-axis and position on the *y*-axis. Label each axis with the appropriate SI units. This graph tells you the position of the car at any time. Describe the shape of the graph.

C. How far did the car travel in each 1.0 s time interval (2.0–3.0 s, 3.0–4.0 s, 4.0–5.0 s, etc.)? For example, to find the distance traveled in the 2.0–3.0 s time interval, subtract the car's position at 2.0 s from the car's position at 3.0 s, and record this value in your notebook. Repeat to find the change in position for each time interval.

D. Predict the position of the car at 12.0 s. Explain your prediction.

E. Use your answers from C to make a graph with time on the *x*-axis and change in position on the *y*-axis. Label each axis with the appropriate SI units. This graph tells you the distance traveled by the car in each time interval. Describe the shape of this graph.

F. Compare the graphs you made in parts B and E. What similarities are there between these two graphs?

Moving at an increasing speed

Procedure

9. Support one end of the track 2 cm–3 cm above the floor with clay as shown. Secure the track so that it does not move. The base of the track should rest on the floor. Place a block of wood on the floor against the base of the ramp. Mark a point near the top of the track with masking tape, and label it "starting point."

10. Place the ball at the starting point. Hold the ball in place with a ruler.

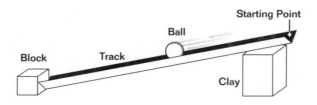

11. To release the ball, rapidly swing the ruler out of the way. Start the stopwatch the instant the ball is released. The ball will roll down the track.

12. Stop the stopwatch when the ball reaches the base of the track.

13. Repeat steps 10–12. Adjust the angle of the track for each trial until you find a position at which it takes the ball slightly longer than 5.0 s to travel from the starting point to the bottom of the track.

14. When the track is secured in position at the determined angle, place the ball at the starting point. Hold the ball in place with a ruler. To release the ball, rapidly swing the ruler out of the way. Start the stopwatch the instant the ball is released.

15. After 4.0 s, mark the position of the ball with masking tape. Label it "4.0 s."

16. Repeat step 14, but mark the position of the ball after 3.0 s of travel. Label the tape "3.0 s."

17. Repeat step 14, but mark the position of the ball after 2.0 s of travel. Label the tape "2.0 s."

18. Measure the exact distance from the starting point to each position marked with tape.

19. For each position, record the distance and time in your notebook, using the appropriate SI units. Make sure to record all measured digits plus one estimated digit.

Analysis

G. Did the ball speed up or slow down as it traveled, or did it maintain the same speed? How can you tell?

H. Make a graph of your data with time on the *x*-axis and position on the *y*-axis. Label each axis with the appropriate SI units. This graph tells you the position of the ball at any time. What shape does the graph have?

I. How far did the ball travel in each 1.0 s time interval (2.0–3.0 s, 3.0–4.0 s, 4.0–5.0 s, etc.)? To answer this, find the distance that the ball traveled in each 1.0 s time interval. For example, to find the distance traveled in the 2.0–3.0 s time interval, subtract the ball's position at 2.0 s from the ball's position at 3.0 s, and record this value in your notebook. Repeat to find the change in position for each time interval.

J. Predict the position of the ball at 12.0 s. Explain your prediction.

K. Use your answers from I to make a graph with "time" on the *x*-axis and "change in position" on the *y*-axis. Label each axis with the appropriate SI units.

L. Compare the shape of the graphs you made in parts H and B. What differences are there between the graphs?

Race-Car Construction

U.S. RACING ASSOCIATION
LYNCHBURG, SOUTH CAROLINA

September 27, 1999

Mr. Steve Thorpe

1% Inspiration Laboratories
14557 West Post Road
Tempe, Arizona 85289

Dear Mr. Thorpe:

To celebrate our 25th anniversary, we are promoting auto racing this season by having a contest to develop an inexpensive race car. Cash awards and free tickets to the U.S. Racing Association Silver Cup race are going to be awarded in each category to the fastest car that meets the criteria.

The contest will include judging in two categories: cars with motors and motorless cars (cars that move by the force of gravity). The cars that include motors should be powered only by batteries (no fuel) and should travel a displacement of 5.0 m. Motorless cars will need to accelerate to top speed using only a ramp or a similar physical structure and should travel a displacement of 3.0 m. The car may not be pushed, launched, or pulled. If you enter this category, you should also include a complete description of the device used to accelerate the car.

All cars should be composed of scrap materials found around the home. The appearance of the car will not be judged, but contestants should pay careful attention to physical design elements that affect the ability of the car to travel in a straight line at high speeds. Each contest entry should include an analysis of the car's speed, using appropriate SI units accurate to three significant digits. The analysis should average the speeds over three trials, traveling a horizontal distance on a smooth surface, such as tile or a similar surface. The speed must be calculated only on the horizontal path of the car's travel. Each contest entry should use the format of a patent application and include the name of the car. Good luck in the design of your contest entry.

Sincerely,

Billy Joe Greenfield

Billy Joe Greenfield

More information about the design is on page 12.

1% ≈ Inspiration Laboratories

MEMORANDUM

Date: September 28, 1999
To: Development Team
From: Steve Thorpe

This project reminds me of some of the soapbox derbies I entered when I was a kid. This really sounds like fun! The U.S. Racing Association car design contest could result in some great prizes, so we will need to do careful planning.

Before you go into the lab, prepare a plan for the design of the car. Your plan should include a list of materials needed and a diagram of the car. Remember to include all of your testing and development procedures. I have included a newspaper clipping with this memo that may be helpful to your design and setup. Your plan should also include a design of a car that will move in a straight path.

- An easy way to do this is to make sure that the car is stable and that it does not pull to either side. Your design should take into account the size and shape of the car.

- For the car without a motor, take into consideration that the car will begin to slow down at some point along its horizontal path.

- Determine the average velocity your car will travel over three trials, and show your calculations.

I will approve your plan before you start work on your project, so turn it in to me soon. When your car is ready, prepare your report using the format of a patent application. Be sure your report includes all parts of the application, and pay close attention to the number of significant figures throughout the lab. Good luck!

14557 West Post Road • Tempe, Arizona 852

See next page for safety requirements, materials list, and more hints.

continued

MATERIALS

ITEM	QTY.
✔ 1.5 V–3.0 V dc motor	1
✔ 15 cm insulated wire	1
✔ AA batteries	2
✔ aluminum sheet	
✔ bamboo skewers	2
✔ drinking straws	2
✔ glue	
✔ large rubber bands	2
✔ masking tape	
✔ meterstick	
✔ plastic film-canister lid	3
✔ scissors	1
✔ small rubber bands	4
✔ stopwatch	1
✔ support stand and clamps	2
✔ table clamp	
✔ tongue depressors	5
✔ inclined plane	1

SAFETY

- Wear eye protection and perform this experiment in a clear area.

- Cut carefully, and be aware of those around you. When working with a knife, do not draw it toward you. After using a sharp tool, cover it with its protective sheath and return it to a safe place. Sharp objects can cause serious injury.

Coaster cars gravitate to a winning speed

In an event that combines elements of automobile racing and downhill sledding, coaster cars zip down a hill under the pull of gravity to pick up speed for the timed run on the flat surface of the track. Cars that win tend to be heavy, narrow, and low to the ground.

Races will be held today at Coaster Lanes. The track measures 50 meters from the starting line at the bottom of the hill to the finish line. The slope of the hill is 20 degrees.

Manuel Sanchez, last year's winner, explains that there are many tricks to

building a successful coaster car. "Wheel alignment is important in making sure that the car will move in a straight path," he says, "Also, knowing how to distribute the mass is critical to building a winning car. You have to make sure the car does not slow itself down."

Vector Treasure Hunt

SAFETY

• Review the lab safety guidelines. Always follow correct procedures in the lab.

MATERIALS

✔ meterstick or trundle wheel
✔ index cards

OBJECTIVES

• Create a series of directions that lead to a specific object.
• Follow directions to locate a specific object.
• Develop a standard notation for writing direction symbols.
• Generate a scale map.

Giving directions

Procedure

1. In this lab, you will select a large, fixed object at your school and use standard physics notation to direct other students to the object. Your teacher will define the starting point and the physical boundaries for this activity. Select an object within the boundaries; the object you choose should be large and obvious, and it should be fixed in place so that other students will be able to find it by following your directions.

2. Plot out a course from the starting point to the chosen object. Remember to work quietly and to avoid disrupting classes and school traffic. Use a meterstick or trundle wheel to measure the distances along the course. Alternatively, you may measure your pace in meters and use your pace to count out the distance for each part of the course. Convert your pace to meters before recording the values for each distance.

3. You will break up the course into 15 different segments, and you will write each separate segment as a distance and a direction on an index card. Each card must contain a complete description of that segment, including the magnitude of the distance in meters and the direction. The direction must be specified using only these terms: north, south, east, west, up, and down. Your teacher will tell you where north is located for the purposes of this lab.

4. Keep in mind that the cards may be used to describe the most direct path from the starting point to the object, broken up into 15 segments, or they may describe a complicated path with many changes of direction.

5. When you have completed 15 cards that give an accurate description of a path between the starting point and the chosen object, write your name on an index card, and place the card on top of the 15 cards. On a separate piece of paper,

write your name and a description of the object you chose, including a description of its location. Give this paper and your deck of direction cards to your teacher. Your teacher will keep the paper with the name of the object until the end of the lab.

Analysis

A. Do your cards describe the straight-line path to the object divided into 15 parts, or do they describe a winding path to the object?

B. Is the path described by your cards the same length or longer than the straight-line path to the object? Can your cards be used to determine the straight-line path? Explain.

C. What was the most difficult part of plotting the path to the object?

D. Are you confident that another group will be able to find the object using your direction cards? Explain why or why not.

E. Would another group be able to find the object using your direction cards if your cards were placed out of order? Explain your answer.

Following directions

Procedure

6. When you turn in your cards, your teacher will shuffle them well and give the shuffled cards to another lab group. You will receive a shuffled deck of direction cards made by another group.

7. Devise a plan to use the directions on the cards you have been given to find the object chosen by the other group, then attempt to find the object.

8. When you find the object, go back through the cards to make sure you have correctly identified the object selected by the other group.

9. When you are sure that you have found the correct object, report your results to your teacher. Your teacher will confirm whether you have correctly identified the object. If not, review the cards and try again.

Analysis

F. Did shuffling the deck make it more difficult for you to locate the object? Explain why or why not.

G. Would you be able to place the cards in their original order? Explain why or why not.

H. Did you find the object described by the other group's cards? If not, explain what happened.

I. Explain the method you used to find the object, and include any tricks you discovered while you were working.

J. Was the other group able to correctly identify the object described by your direction cards?

Mapping the course

Procedure

10. In this section of the exercise, you will use the directions on a set of 15 cards to draw a map of the path from the starting point to the object. You will generate a map of the complete set of directions you used to find the object.

11. You will make the map by drawing each direction indicated on a card as an arrow. The arrow will be drawn to scale to represent the length in meters and it will point in the direction specified on the card. In a scale drawing such as this, it is important for all the objects in the drawing to have the same size relationship as the actual objects. For example, the arrow representing 2.0 m will be drawn twice as long as an arrow representing 1.0 m.

12. Draw the first arrow so that its tail is at the starting point, the point of the arrow is pointing in the direction specified on the card, and the length of the arrow represents the distance on the card.

13. Draw the second arrow on your map so that its tail starts at the point of the first arrow. The second arrow should also point in the direction specified by the card, and its length should represent the distance on the card.

14. Continue through the entire set of 15 cards. Draw the arrows tip-to-tail so that each arrow begins where the preceding one ends.

15. Make sure that the map is very neat. Include a legend, or key, that gives the directions and defines the scale of the map. You may wish to indicate specific landmarks, such as rooms or doors.

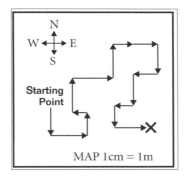

Analysis

K. Does the map accurately reflect the path you took to find the object? If not, explain any differences.

L. Explain how shuffling the cards affected the way you represented the directions from the starting point to the object. Use examples from your map to support your answer.

M. Based on this exercise, describe the most efficient method of using the set of direction cards to locate the object. Would this work for any set of directions? Explain why or why not.

The Path of a Human Cannonball

THE AMAZING LASLO CIRCUS
KITTANNING, PA

October 11, 1999

Dr. Wes Graham
1% Inspiration Laboratories
14557 West Post Road
Tempe, Arizona 85289

Dear Dr. Graham:

I spoke with you recently regarding our new "Human Cannonball" act, in which our daredevil, Clem, will be launched from a cannon into a net.

Our obvious problem is, how can we predict where to place the net? Using a portable radar gun, we've measured Clem's speed as he leaves the cannon. The net is strong enough to withstand the force of Clem's impact. In the first act, we plan to launch Clem so that he lands at the same horizontal level from which he was launched. For extra thrills, we will eventually mount a flaming ring at the highest point of his path so he can fly through the ring. Later in the show, Clem will be launched from a high platform and will land on a net placed far below the platform. For both acts, Clem's launch speed will be known, and we will determine the initial angle of launch and the placement of the net and ring based on your report.

Clem wears a special nylon suit and helmet that reduce air resistance significantly, so this should not be a problem. Also, I'm not sure if it matters, but Clem is 1.7 m tall, and he weighs 175 pounds.

Our tour starts in three months, so time is a critical factor here. On the other hand, a man's life is at stake, so accuracy is more important. Thank you very much for your time.

Respectfully,

John Lerner

John Lerner

Diagrams of the human cannonball act are on page 18.

1% Inspiration Laboratories

MEMORANDUM

Date: October 15, 1999
To: Research and Development Team
From: Wes Graham

You probably remember that I mentioned this contract at the last departmental meeting. Attached are copies of the letter and basic design specs, along with a list of relevant materials in stock. Start working up a plan so you can go into the lab as soon as possible. Work in SI, and keep track of significant figures. Present your plan to me before you start work. Make sure your plan includes the equipment you'll need and the measurements you are planning to take. You should also figure out what equations you'll need to determine where the net and ring should be placed for each part of the act.

As far as I can tell, this looks like a simple projectile-motion problem. Develop the equations and models to predict the maximum vertical and horizontal displacements at different angles. That will allow us to make recommendations based on our tests. Let's perform tests for launching at 20°, 40°, and 60°. For each angle, we need to recommend the placement of the ring and of the net.

For each part of the act, I think we should provide a set of equations and a working model of the act. We want to make sure that the equations will give the correct placement of the net for any angle they start with, given the initial speed. Pay special attention to answering the following questions:

- Exactly where should the net be placed?

- Where should the center of the flaming hoop be placed?

As I said at the meeting, it has been a great year at this company thanks to all of you.

14557 West Post Road • Tempe, Arizona 852

See next page for safety requirements, materials list, and more hints.

MATERIALS

ITEM	QTY.
✔ adhesive tape	1 roll
✔ ball launcher and ball	1
✔ carbon paper	1 sheet
✔ cardboard box	1
✔ clamps	3
✔ clay	200 g
✔ cloth towel	1
✔ lattice rod	1
✔ meterstick	1
✔ metric ruler	1
✔ photogate timing system	1
✔ plumb bob and line	1
✔ protractor	1
✔ support stand and ring	2
✔ white paper	1 sheet

SAFETY

- Wear eye protection, and perform this experiment in a clear area. Falling or dropped masses can cause serious injury.

Act 1

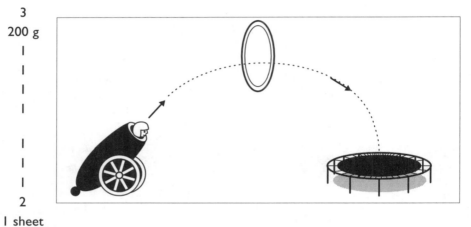

Act 2

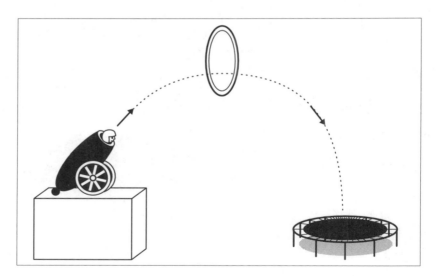

Discovering Newton's Laws

SAFETY

- Perform this experiment in a clear area. Falling or dropped masses can cause serious injury.

- Tie back long hair, secure loose clothing, and remove loose jewelry to prevent their getting caught in moving or rotating parts.

MATERIALS

✔ 3 masses, 1 kg each

✔ beaker

✔ coin, such as a quarter

✔ cord

✔ dynamics cart

✔ dynamics cart with spring mechanism

✔ human-figure toy or doll

✔ index card

✔ paper towels

✔ rubber band

✔ set of masses, 20 g–100 g

✔ stopwatch

✔ track with pulley and car

✔ water

OBJECTIVES

- Explore the factors that cause a change in motion of an object.
- Determine the effect of mass on an object's acceleration.
- Investigate the acceleration of two objects acting on one another.

An object at rest

Procedure

1. Carefully fill the beaker about half-full with water. Wipe the lip and the outside of the beaker with a paper towel.

2. Place an index card on top of the beaker so that the card covers the opening of the beaker. Place the quarter on top of the card.

3. Remove the index card by pulling it quickly away. Make sure you pull the card perfectly horizontally.

Index Card

Coin

Beaker
of Water

Analysis

A. What happened to the coin when the card was pulled out from underneath?

B. Is this what you expected to happen? Explain why or why not.

C. What would happen to the coin if the card were pulled out very slowly? Try it, and compare your results.

An object in motion

Procedure

4. Choose a location where you can push a dynamics cart so that it rolls for a distance without hitting any obstacles or obstructing traffic and then hits a wall or other hard surface.

5. Place the toy or doll on the cart, and place the cart about 0.5 m away from the wall.

6. Push the cart and doll forward so that they run into the wall. Observe what happens to the doll when the cart hits the wall.

7. Place the cart at the same starting place, about 0.5 m away from the wall. Return the doll to the cart, and use a rubber band to hold the doll securely in the cart.

8. Push the cart and doll forward so that they run into the wall. Observe what happens to the doll when the cart hits the wall.

9. When you are finished, return the cart to the table or storage place. Do not leave the cart on the floor.

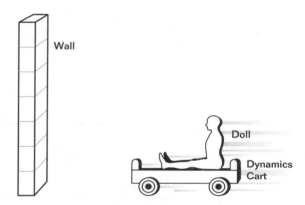

Analysis

D. What happened to the unsecured doll when the cart hit the wall?

E. What happened to the doll secured with the rubber band when the cart hit the wall?

F. How did the rubber band change the result of the experiment? Explain why this happened.

G. Compare the experiment with the doll and cart with the experiment with the card and coin. Explain how the results of the two are similar.

Newton's second law

Procedure

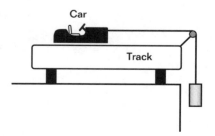

10. Perform this part of the lab using an air track and car or a dynamics track and car. Place the car on one end of the track with the pulley securely clamped to the other end of the track.

11. Securely attach one end of a cord to the car and the other end to a small mass. Thread the cord through and over the pulley wheel at the end of the air track or dynamics track. The car should be held securely in place at the opposite end of the track.

12. Make sure that the mass will be able to fall about 1 m without hitting any obstacles. If you are using the air track, turn on the air track and release the car at the same moment. If you are using the dynamics track, release the car. The mass will fall straight down, and the car will move along the track. Be ready to catch the car when it reaches the end of the track.

13. While the car is moving, make careful observations. Try to determine whether the car moves with constant velocity or whether it accelerates.

14. Replace the mass with another mass, and repeat steps 10–13. Carefully observe the motion of the car.

15. Repeat several times using different masses. Do not exceed 300 g. As you change the mass, watch the motion of the car for observable patterns.

Analysis

H. What caused the car to start moving?

I. Did the car move with a constant velocity, or was it accelerating?

J. How did the size of the falling mass affect the motion of the car? Explain.

Newton's third law

Procedure

16. Set up two dynamics carts as shown. Choose a location where each cart will be able to move at least 1.0 m on a smooth horizontal surface away from obstacles and traffic. Compress the spring mechanism and place the carts so that they are touching, as shown.

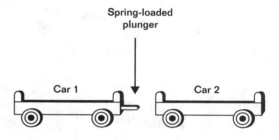

17. Quickly release the spring, and observe the two carts. If you are working on a lab table, do not allow the carts to fall off the table.

18. Return the carts to the original position, and compress the spring mechanism. Add a 1 kg mass to the cart with the spring.

19. Quickly release the spring, and observe the two carts.

20. Return the carts to the original position, and compress the spring mechanism. Add another 1 kg mass to the cart with the spring. Release the spring, and observe the two carts.

21. Return the carts to the original position, and compress the spring. Add a 1 kg mass to the second cart so that the mass on the first cart is twice the mass on the second cart. Release the spring, and observe the two carts.

Analysis

K. What happened to the two carts when the spring was released?

L. Compare the motion of the carts for each trial. Describe the motion in terms of the carts' acceleration from rest when the carts have equal mass (no masses added), one cart has 1 kg mass added, one cart has 2 kg mass added, and when one cart has 2 kg mass added and the other cart has 1 kg mass added.

M. What is the relationship between the mass of a cart and its acceleration when the spring is released?

Friction: Testing Materials

1% Inspiration Laboratories

MEMORANDUM

To: Dr. Jan Ingensen, Research and Development
From: L. Morales, Inventory Manager
Date: November 11, 1999

In order to comply with the new labeling regulations, we have been going through the materials supply room and replacing labels that no longer meet the required specifications. A recent inventory of the materials supply room has revealed a large surplus of untested materials. Many of these are surface-coating materials used to reduce friction between surfaces, or in some cases, to increase it.

In order to update the labels on these products, we need to ascertain their functions. With our new inventory system, we will be labeling these items based on the coefficient of friction. I have included a list of the un-tested materials in the storeroom. Please test each of these materials, and let me know the results by the end of next week. Be sure to give me all the documentation.

That's all for now. Thanks a lot.

14557 West Post Road • Tempe, Arizona 85289

The list of materials to be tested is on page 24.

1% Inspiration Laboratories

MEMORANDUM

Date: November 12, 1999
To: Research and Development Team
From: Jan Ingensen

I've looked over the list sent down from Inventory. With the recent hiring boom in the company, I think we have enough people to get these results in time to make the deadline for the new labels. We may even have freedom to do extra tests on these for future reference.

Look over the list I've included with this memo. It gives all the materials that need testing, and I've added the equipment we have available for performing the tests. Some of these materials have been used in the manufacture of nonslip feet (for appliances, bathtubs, etc.), while others have been used to reduce friction to aid in pushing large objects. Come up with a plan to analyze these materials for their relative value to reduce or increase friction. Remember to get my approval for your plan before you go into the lab to begin testing.

I think we should test each material against the same material so that we can compare the coefficients of friction. Make sure to perform the same tests on all the items on the list. Find the coefficients of static and kinetic friction to two significant figures. When you have your results, rank them in order of the coefficients of friction for each test, and be sure to distinguish between static and kinetic friction. Give me a full report detailing the tests you performed and your results. I would also be interested to see whether the rank according to the coefficients of kinetic friction is the same as the rank according to static friction.

14557 West Post Road • Tempe, Arizona 852

See next page for safety requirements, materials list, and more hints.

MATERIALS

ITEM	QTY.
✔ balance	1
✔ cork board	1 sheet
✔ force meters	2
✔ linoleum	1 sheet
✔ masking tape	1 roll
✔ sandpaper	1 sheet
✔ set of masses	1
✔ unidentified materials	1 box
✔ wooden friction block with hook	1

SAFETY

- Perform this experiment in a clear area. Falling or dropped masses can cause serious injury.

- Tie back long hair, secure loose clothing, and remove loose jewelry to prevent their getting caught in moving or rotating parts.

Keep in mind that the coefficient of friction describes a relationship between two surfaces. Your reports should include a complete description of both surfaces in each test. If there is time, perform all the tests against a second material to see if the ranking according to the coefficients of friction is the same regardless of what material you test against.

Make sure that you keep records of all data and measurements used to find the coefficient of friction. Because the coefficient of friction is a ratio of measured or calculated forces, it is important that you carefully document all your measurements.

HRW material copyrighted under notice appearing earlier in this book.

Exploring Work and Energy

MATERIALS

✔ clamps
✔ cord, 1.00 m
✔ force meter
✔ inclined plane
✔ masking tape
✔ meterstick
✔ set of hooked masses
✔ stopwatch

SAFETY

- Set up the apparatus, and attach all masses securely. Perform this experiment in a clear area. Swinging or dropped masses can cause serious injury.

- Tie back long hair, secure loose clothing, and remove loose jewelry to prevent their being caught in moving or rotating parts.

OBJECTIVES

- Measure the force required to move a mass over a certain distance using different methods.
- Compare the force required to move different masses over different time intervals.

Pulling masses

Procedure

1. At one edge of the tabletop, place a tape mark to represent a starting point. From this mark, measure exactly 0.50 m and 1.00 m. Place a tape mark at each measured distance.

2. Securely attach the 1 kg mass to one end of the cord and the force meter to the other end. The force meter will measure the force required to move the mass through different displacements.

3. Place the mass on the table at the starting point. Hold the force meter parallel to the tabletop so that the cord is taut between the force meter and the mass. Carefully pull the mass at a constant speed along the surface of the table to the 0.50 m mark (this may require some practice). As you pull, observe the force measured on the force meter.

4. Record the force and distance in your notebook using the appropriate SI units.

5. Repeat steps 3 and 4 for a distance of 1.00 m.

6. Repeat steps 3, 4, and 5 with a 0.2 kg mass.

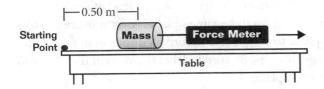

Analysis

A. Did you exert the same force on the 1 kg mass as you did on the 0.2 kg mass to move them an equal distance?

B. Did it require more force to move the mass 1.00 m than to move the same mass 0.50 m?

C. What force did you pull against?

Lifting masses

Procedure

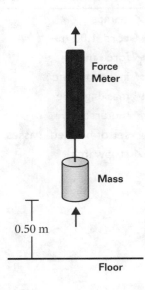

7. Using masking tape, secure a meterstick vertically against the wall with the 0.00 m end on the floor.

8. Securely attach the 1 kg mass to one end of the cord and the force meter to the other end.

9. Place the mass on the floor beside the meterstick. Hold the force meter parallel to the wall so that the cord is taut between the force meter and the mass. Carefully lift the mass vertically at a constant speed to the 0.50 m mark on the meterstick. Be sure that the mass does not touch the wall during any part of the process. As you lift, observe the force measured on the force meter. Be careful not to drop the mass.

10. Record the force and distance in your notebook using the appropriate SI units.

11. Repeat steps 9 and 10 for a vertical distance of 0.25 m.

12. Replace the 1 kg mass with the 0.2 kg mass, and repeat steps 9, 10, and 11.

Analysis

D. Did you exert the same force on the 1 kg mass as you did on the 0.2 kg mass to move them an equal distance?

E. Did it require more force to lift the mass 0.50 m than was required to lift the same mass 0.25 m?

F. What force did you lift against?

G. Did it require a different force to *lift* a mass than it did to *pull* the same mass across the table an equal distance?

Displacing masses using an inclined plane

Procedure

13. Carefully clamp an inclined plane to the tabletop so that the base of the inclined plane rests on the floor. Make sure the inclined plane is in a location where it will not obstruct traffic or block aisles or exits.

14. Measure vertical distances of 0.25 m and 0.50 m above the level of the floor. Use masking tape to mark each level on the inclined plane. Also measure the distance along the inclined plane to each mark. Record all distances in your notebook using the appropriate SI units. Be sure to label the vertical distance and the distance along the inclined plane.

15. Attach the 1 kg mass to the lower end of the cord and the force meter to the other end.

16. Place the mass at the base of the inclined plane. Hold the force meter parallel to the inclined plane so that the cord is taut between the force meter and the mass. Carefully pull the force meter at a constant speed parallel to the surface of the inclined plane until the mass has reached the vertical 0.50 m mark on the inclined plane. As you pull, observe the force measured on the force meter.

17. Using the appropriate SI units, record the force and distance in your notebook.

18. Repeat steps 16 and 17 for a vertical distance of 0.25 m.

19. Repeat steps 16, 17, and 18 for the 0.2 kg mass.

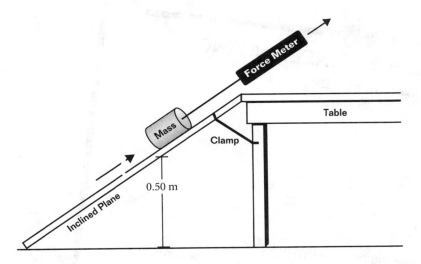

Analysis

H. Did you exert the same force on the 1 kg mass as you did on the 0.2 kg mass to move them an equal distance?

I. Did it require more force to lift the same mass 0.50 m along the inclined plane as it did to lift it 0.25 m?

J. What forces did you pull against?

K. Compare the force required to lift a mass using an inclined plane with the force required to lift the same mass to the same vertical displacement using only the force meter. Why are the values different?

L. How can you adjust the inclined plane so that moving the mass through the same vertical displacement requires less force?

Bungee Jumping: Energy

NISKAYUNA HIGH ENGINEERING INC.

SCHENECTADY, NY 12309

December 10, 1999

Dr. John R. Kanga
1% Inspiration Laboratories
14557 West Post Road
Tempe, Arizona 85289

Dear Dr. Kanga:

Since the inception of bungee jumping nearly ten years ago, the development of equipment for this sport has been stagnant. Sheathed shock cords have been the only apparatus used in this activity. These cords lend no creativity in design for either application or appearance. As a result, there has been a decline in interest in the sport and, in turn, drops in the ride fees our clients can charge. It is our goal to promote new interest in the sport and to bolster sales by designing upgraded equipment for owners of current bungee-jumping operations.

We are seeking a new design for a bungee cord that will safely bring a diver to a smooth halt at the bottom of the flight. The new design should incorporate the use of our newly developed elastic bands and braided cords. Included in this mailing is the equipment that we have available for use in designing the new bungee cord. You must not include any other devices in the design, and you must use all the equipment enclosed.

To use humans in such experimentation is unwise and to perform a full size operation would not be practical, so a scaled-down model of the design is appropriate. Primarily, we must be certain that the diver would be safe. As a result, we require data from tests of your design. Your design, along with designs from other engineering firms, will be tested by our firm only once. A contract will be offered to the firm whose bungee cord stops the diver closest to the floor without touching the floor.

Sincerely,

Dr. Sun Nguyen

Dr. Sun Nguyen

More information about the design is on page 30.

1% Inspiration Laboratories

MEMORANDUM

Date: December 13, 1999
To: Development Team
From: Dr. John R. Kanga

The bungee-cord-design request from Dr. Sun Nguyen could lead to a big contract, so we will need to do careful planning. Before you go into the lab, prepare a plan for the design of the bungee cord. Your plan should include a list of materials needed and a diagram of the experimental setup. You will also need a data table for the mass, cord length, expected length of fall, and the spring constant of the elastic bands included in this kit. Remember to document all of your testing and development procedures in your lab notebook. I have included with this memo a newspaper clipping that may be helpful. Your plan should also include the following:

- a bungee-cord design that uses only the braided cord and the elastic bands provided in the kit. This means that you will need to justify the choice of bungee-cord length. Since this length depends on how much the elastic bands will stretch, you should also use equations to demonstrate how you will determine the spring constant of the elastic bands provided in the kit.

- recommendations of ways to bring the diver to a smooth halt. It may be helpful to consider the principle of conservation of energy in this situation.

I must approve your plan before you start work in the lab, so turn it in to me soon. You will receive the kit of braided cords and elastic bands when I approve your plan. After your work in the lab, prepare your report using the format of a patent application. Be sure your report includes all eight parts of the application. Good luck!

14557 West Post Road • Tempe, Arizona 852

See next page for safety requirements, materials list, and more hints.

continued

MATERIALS

ITEM	QTY.
✔ clamps	3
✔ heavy cardboard	10 cm × 10 cm
✔ Hooke's law apparatus	
✔ meterstick	1
✔ set of slotted masses	1
✔ slotted mass holder	1
✔ suspension clamp	1

KIT INCLUDES:

ITEM	QTY.
✔ braided cords	1.5 m–2.0 m
✔ elastic bands	2 or 3
✔ hooked masses	0.2 kg, 0.25 kg or 0.5 kg

Plunge with a Bungee

Although bungee jumping has been a craze for almost a decade, many people are wondering just how safe such a plummet can be. A harnessed person secured to one end of a long elastic bungee cord attaches the other end of the cord to a high precipice, such as a bridge or a cliff. After summoning the courage, they plunge and are rewarded with the exhilarating free-fall acceleration of their body toward the ground. When the diver has fallen the length of the cord, the cord gives a little, much as a spring does. So it's important that designers know *exactly* how much the cord will give when they determine the length of the cord. Designers must also take into account the range of weights of different people. Although the fall is fun for many divers, some have complained about the jolt experienced at the end of the ride. When the cord cannot expand any further, it yanks the diver back up away from the ground—causing the diver to fall again and experience another, less harsh jolt. The entire experience is much like that of a bouncing ball.

Circular Motion

SAFETY

- Wear eye protection and perform this experiment in a clear area away from electrical equipment or outlets. Clean up any spilled or splashed water immediately.

- The bands will break if they spin too quickly or in a figure 8. If the elastic bands break during the experiment, serious injury could result.

- Tie back long hair, secure loose clothing, and remove loose jewelry to prevent their being caught in moving or rotating parts.

- Rotating or swinging masses can cause injury.

MATERIALS

- ✔ 8 elastic bands, ⅛ in. wide
- ✔ balance
- ✔ meterstick
- ✔ plastic bottle marked at the 150 mL level
- ✔ 14 oz. plastic drinking cup with three equally spaced holes below the rim
- ✔ stopwatch

OBJECTIVES

- Distinguish between forces required to hold a variety of masses in a horizontal circular path moving at several speeds.
- Compare the circular motion of masses to the linear motion of masses.
- Discover the relationship between mass, speed, and the force that maintains circular motion.

Slow circular motion with a mass

Procedure

1. Push an elastic band through a hole below the rim of the plastic cup. Loop the band through itself as shown. This action should form a type of knot about the rim of the glass. Secure the knot tightly.

2. Repeat step 1 for each hole in the plastic cup.

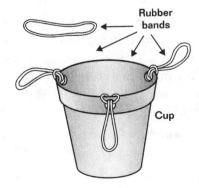

Rubber bands

Cup

3. Pull another elastic band through each knotted band on the cup's rim so that all three bands on the rim of the cup simultaneously loop around the fourth band. Make a knot similar to the one you made in step 1. This will knot all bands together and create a fourth band loop.

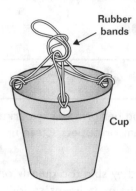

Rubber bands

Cup

4. Loop another elastic band through the fourth band in the same way that you did in step 1. Then loop three more elastic bands end to end in a chain to lengthen the device. This device is referred to as a *cupsling*.

Rubber bands

Cup

5. Carefully measure 150 mL of water into the plastic cup. Make sure that no water spills.

6. Place the cupsling on a balance, and record its mass using the appropriate SI units. Make sure to record all measurements to the precision of your balance.

7. Make sure the area is clear of obstacles, and warn other students that you are beginning your experiment. Holding the bands securely, slowly spin the full cupsling about you in a full circle. Slightly increase the speed until you can spin it so that the cup moves in a horizontal circle. Try to see how slowly you can spin the cupsling and still consistently maintain a horizontal circle. Be careful not to spill or splash any water.

8. With the stopwatch, a partner should time the 10 complete circles of the cup as you swing it *slowly* around in a *horizontal* circle.

9. A partner should use the meterstick to estimate the radius of the cup's horizontal path at this speed. Get as precise an estimate as possible. Always be aware of the position of the cupsling.

10. Using the appropriate SI units, record the radius of the circle and the total time it took to complete 10 horizontal circles of the cup in your notebook. Make sure to record all measured digits plus one estimated digit.

Analysis

A. Did you need to exert a force on the elastic band to start spinning the cupsling from rest?

B. Did you need to continue exerting a force on the elastic band to keep it spinning at a constant speed? How did you know?

C. When the cupsling moved in a circle, it was changing direction all the time. What caused the cupsling to change direction?

D. When the cupsling moved in a circle at a constant speed, did it accelerate? Explain your answer.

E. Where do you think the cup would go if the band were released while the cup was spinning?

F. What happened to the length of the elastic band as you increased the force to spin the cupsling in a horizontal circle?

G. How long did it take for the cup to complete one circle?

Circular motion with a mass

Procedure

11. Place the cupsling with 150 mL of water in the cup on a balance. Record its mass using the appropriate SI units. Make sure to record all measurements to the precision of your balance.

12. Holding the bands securely, slowly spin the full cupsling about you in a complete circle. Slightly increase the speed until you can spin it so that the cup moves in a horizontal circle. Spin the cupsling faster than you did in step 7 but not so fast that the bands will break. Remember to consistently maintain a horizontal circle throughout this experiment. Be careful not to spill or splash any water.

13. With the stopwatch, a partner should time the 10 complete horizontal circles of the cup.

14. Using the meterstick, a partner should estimate the radius of the cup's horizontal path at this speed. Get as precise an estimate as possible. Always be aware of the position of the cupsling.

15. Using the appropriate SI units, record the radius of the circle and the total time it took to complete 10 horizontal circles of the cup in your notebook. Make sure to record all measured digits plus one estimated digit.

Analysis

H. What happened to the length of the elastic band as the speed increased?

I. What happened to the force on the elastic band as the speed increased?

J. How long did it take for the cup to complete one circle?

Circular motion with an increased mass

Procedure

16. Place the cupsling with a total of 300 mL of water on a balance. Record its mass using the appropriate SI units. Make sure to record all measurements to the precision of your balance.

17. Make sure the area is clear of obstacles, and warn other students that you are beginning your experiment. Holding the bands securely, slowly spin the full cupsling about you in a full circle. Slightly increase the speed until you can spin it so that the cup moves in a horizontal circle. Try to see how slowly you can spin the cupsling and still consistently maintain a horizontal circle. Be careful not to spill or splash any water.

18. With the stopwatch, a partner should time the 10 complete circles of the cup as you sling it *slowly* around in a *horizontal* circle.

19. Using the meterstick, a partner should estimate the radius of the cup's horizontal path at this speed. Get as precise an estimate as possible. Always be aware of the position of the cupsling.

20. Using the appropriate SI units, record the radius of the circle and the total time it took to complete 10 horizontal circles of the cup in your notebook. Make sure to record all measured digits plus one estimated digit.

Analysis

K. What happened to the length of the elastic band when you increased the mass in the cup?

L. How did the increase in mass affect the force on the elastic band?

M. If a mass moves in a straight line and more mass is added, does the inertia increase, decrease, or stay the same?

N. Do you think that the same thing happens to a body in circular motion? Explain.

O. How long did it take for the cup to complete one circle?

Pre-Chapter Exploration 8

Discovery Lab

Torque and Center of Mass

MATERIALS

- ✔ 1.25 cm diameter dowel rod, 0.5 m long
- ✔ 2 frozen-juice cans and lids
- ✔ 15 mm bolt, 5 cm long
- ✔ 15 mm nut
- ✔ 15 mm washer
- ✔ adjustable wrench
- ✔ apple
- ✔ clay
- ✔ cord, 1.00 m
- ✔ force meter
- ✔ masking tape
- ✔ masses, 20 g, 50 g, and 100 g
- ✔ plastic cup with handle
- ✔ support stand with clamps
- ✔ table clamp
- ✔ wooden plank with a drilled 15 mm hole

SAFETY

- Attach masses securely. Swinging or dropped masses can cause serious injury.
- Tie back long hair, secure loose clothing, and remove loose jewelry to prevent their being caught in moving or rotating parts.

OBJECTIVES

- Discover what factors cause an object to rotate when a force is applied.
- Construct a model of the human arm, and examine the role of forces and rotation in its function.
- Locate the point about which an object that is free to rotate will pivot.

Rotational force and a wrench

Procedure

1. Secure a table clamp to the edge of the table. Use the table clamp to hold the wooden plank vertically. The wooden plank should not move when force is applied to it. Put the bolt through the hole in the wooden plank. Place the washer and the nut on the other side of the plank so that the bolt goes through the hole of the washer and then through the hole of the nut. Adjust the wrench so that it fits snugly around the nut.

2. Firmly grip the tail of the wrench, and use the wrench to tighten the nut. Make sure that the wrench does not slip and that your fingers do not get pinched or jammed.

3. Firmly grip the *head* of the wrench, and try to loosen the nut.

4. Firmly grip the tail of the wrench, and use the wrench to tighten the nut.

5. Firmly grip the *tail* of the wrench, and try to loosen the nut.

Analysis

A. Describe the force that causes the nut to turn when you tighten it. Draw a diagram of the setup showing the direction of the force as it is applied.

B. If you push the wrench into the bolt rather than rotate it, does anything happen?

C. Around which point does the motion of the nut and the wrench occur?

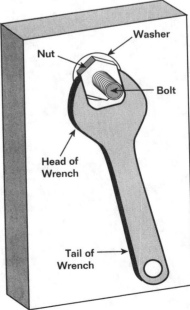

Washer

Nut

Bolt

Head of Wrench

Tail of Wrench

D. What angle between the wrench and the bolt is necessary for you to get the nut to turn the easiest?

E. Was it easier to loosen the nut with your hand by the head or by the tail of the wrench? Was it easier to loosen the nut by applying the force closer to or farther from the bolt?

F. Why do you think the nut stops turning?

Balanced rotational force and the human arm

Procedure

6. Set up the support stand as shown. Place one clamp 20 cm from the top of the stand and another clamp at the bottom of the stand. Make sure both clamps are perpendicular to the support stand.

7. Hook the force meter around the end of the top clamp. There should be at least 4 cm between the force meter and the support stand. Tie a piece of cord into a small loop 3 cm–5 cm in diameter. Hang the loop from the free hook on the other end of the force meter.

8. Thread the dowel rod through the hanging loop and clamp the end of the dowel to the support stand with the lower clamp. Adjust the clamped end of the dowel so that the dowel rod can move freely without falling out of the clamp. To do this, tape a piece of card to one side of the dowel as shown. Pull the cord tightly around the clamp and support rod, and tape the cord securely to the other side of the dowel rod.

9. Tie another piece of cord into a small loop 3.0 cm–5.0 cm in diameter. Hang this loop from the free end of the dowel rod and tape it securely to the dowel rod.

10. Hang a mass of 20.0 g from this loop.

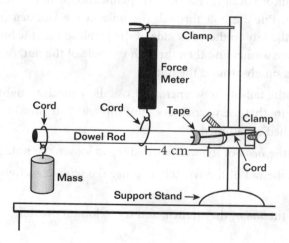

11. Adjust the entire setup so that the force meter and the mass are parallel to each other but perpendicular to the dowel rod. There should be about 4 cm along the dowel rod between the force meter and the clamp, as shown.

12. Observe the force measured on the force meter. Record the mass and the force in your lab notebook. Be sure to use appropriate SI units.

13. Add a 50.0 g mass to the 20.0 g mass, and repeat steps 11 and 12.

14. Add a 100.0 g mass to the 50.0 g and 20.0 g masses, and repeat steps 11 and 12.

15. Measure the distance to the masses from the clamp and the distance from the spring scale to the clamp. Record these values in your lab notebook using appropriate SI units. Be sure to record all digits plus one estimated digit.

Analysis

G. Look at your arm. What part of your arm is represented by the force meter in the model?

H. What part of your arm is represented by the dowel rod in the model?

I. What part of your arm is comparable to the clamp in the model?

J. About what point in the arm model does the rotation occur? What part of your arm does this correspond to?

K. What happens to the force meter when a mass is placed on the loop? Explain.

L. When does the dowel rod move?

M. At what point on the dowel rod is the force applied by the mass? Draw a diagram of the setup showing the direction of this force as it is applied.

N. What produces a force to balance the force due to a hanging mass and prevent the dowel from dropping downward? Draw the direction of the applied force on the diagram of the setup.

O. In which directions are the two forces exerted on the dowel rod?

P. What must happen with these two forces for the dowel rod to not move?

Q. From your observations of the model arm, do the actual forces always cancel each other out?

The pivot point of a freely rotating object

16. Pack clay firmly 1 cm–2 cm deep in one end of an empty frozen-juice can. Seal a lid on this can using masking tape.

17. Tie a 0.50 m cord securely around any part of the can so the can is free to swing. Suspend the can from a support stand and clamp.

18. Draw a line vertically from the suspended part of the cord down the side of the can. You may need to hold the can steady as you draw the line. You may use a ruler or a meterstick to guide you.

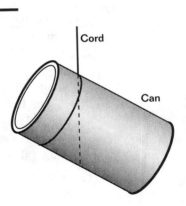

19. Tie the cord around a different part of the can, and suspend the can. Draw a vertical line from the suspended part of the cord down the side of the can. You may need to hold the can steady as you draw the line. You may use a ruler or a meterstick to guide you. Mark the place on the can where the vertical lines meet.

20. Repeat steps 17–19 for another empty frozen-juice can that has been fully packed from top to bottom with clay. Leave no air spaces. Seal a lid on this can using masking tape.

21. Repeat steps 17–19 for an apple on one end of a dowel rod, a pen, and a meterstick.

Analysis

R. Is the point at which the two lines meet the center of the object? Explain.

The Rotating Egg Drop

<u>National Engineering Association</u>
Tempe, Arizona

December 16, 1999

Mr. David Corricello
1% Inspiration Laboratories
14557 West Post Road
Tempe, Arizona 85289

Dear Mr. Corricello:

We are having an Engineering Fair in Wilkshire Mall on December 29 and 30 as part of an effort to inspire both children and adults to consider a career in the engineering industry. We are inviting all of the local engineering firms to set up educational exhibits in the mall. This would be a wonderful opportunity for your company to get some public exposure as well as to help foster awareness of engineering.

We would like to draw a crowd by having all the engineering firms set up an egg-drop exhibit. All egg-drop devices will be dropped from the second floor of the mall to land on the first floor. The object will be to protect the egg from breaking upon impact with the tile floor. To make the project more challenging, we are asking that all companies construct a frame around an egg using only toothpicks and glue. Note that the egg cannot be cooked in any way, nor can it be coated with glue. In addition, rotational motion must be taken into account in the design.

We would like this to be an educational effort as well, so all exhibits must provide a detailed explanation of why the design works. Each presentation should be in the form of a patent application. Each display must include a well-labeled drawing of the device and a sketch of the drop. Labels and captions should be placed below or within every picture.

The registration deadline is December 23. We can accept five entries from each firm. We wish you the best in your egg-drop design.

Sincerely,

Majesh Patel

Majesh Patel

More information about the design is on page 40.

1% Inspiration Laboratories

MEMORANDUM

Date: December 17, 1999
To: Development Team
From: David Corricello

This is a great opportunity to have our work displayed so that people can see it! Before you go into the lab, list the materials needed for the egg-drop device, draw the device's design, and sketch out the drop. Remember to label everything and to provide an informative caption beside each picture. I have jotted down some ideas and am including them with this memo. They might be helpful to your design and setup. You should also use your plan to do the following:

- Explain how Newton's laws of motion and the impulse equation apply to this situation. Use equations in your explanations, and describe how some details of your design influence the magnitude of the variables.

- Describe how rotational motion applies to the design of your egg-drop device.

- Comment on my comparison of the egg-drop device to a weather vane. State whether my ideas are correct, and explain your reasoning.

- Illustrate how the design of your egg-drop device incorporates concepts of rotational forces and torque.

I must approve your plan before you start work on your project, so turn it in to me soon. The five best egg-drop devices will be entered into the exhibit. After your work in the lab, prepare your report using the format of a patent application and include a complete explanation of why your design works. Be sure your report includes all eight parts of the application.

14557 West Post Road • Tempe, Arizona 852⬜

See next page for safety requirements, materials list, and more hints.

MATERIALS

ITEM	QTY.
✔ glue	1 bottle
✔ raw egg	1
✔ toothpicks	1 box

Problem: Orienting the egg-drop device so that I can predict how it will hit the ground.

Is this related to how a weather vane operates?

All weather vanes, which spin about a fixed axis, rotate to face into the wind as the air blows on them. If the weather vane is perpendicular to the wind, the air pushes equally on all parts of the vane. Because the tail of the vane is farther from the axis of rotation, there is more torque per unit area on the tail than on the head (because torque depends on the force and the distance from the rotation axis). So, there is a difference in torque on either side of the axis. This pushes the tail away from the wind and forces the head to face into the wind.

Suppose I construct the device so that it has a toothpick tower on one side: If I drop the device horizontally, the air moves faster and faster as the device falls, and upward forces create a torque on both sides of the center of mass. The tower side is long, so forces on the far end of this side will produce large torque. This will push the tower side back, and the bottom of my egg-drop device should point toward the ground and land first!

HOLT PHYSICS
Discovery Lab

Temperature and Internal Energy

SAFETY

- Perform this experiment in a clear area.
- If a thermometer breaks, notify the teacher immediately.

MATERIALS

✔ 200 g copper shot
✔ 4 plastic-foam cups with lids
✔ balance
✔ hot tap water
✔ ice cubes of uniform size
✔ masking tape
✔ paper towels
✔ plastic container with a 100 mL mark
✔ sharpened pencil
✔ stopwatch
✔ thermometer
✔ weighing paper

OBJECTIVES

- Investigate the phenomenon of energy transfer by heat.

Melting ice cube contest

Procedure

1. Hold an ice cube in your hand so that it melts slower than anyone else's cube in the room.

Analysis

A. How did you hold the ice cube to cause it to melt slowly?

B. Did the heat from your hand influence how fast the ice cube melted?

Energy changes the temperature of copper

Procedure

2. Stack two sets of two plastic-foam cups so that one cup is inside the other cup. Tape each set of stacked cups together.

3. Twist a pencil to carefully bore a hole in the center of the bottom of the first stack of two cups so that the diameter of the hole is the same size as the thermometer. Do not use the thermometer to punch the hole. Make sure that the pencil punctures both cups and that the two holes align.

4. Use the balance to measure out 200 g of copper. Place the copper in the second set of stacked cups.

5. Place the first set of stacked cups upside down on top of the second set of stacked cups so that the rims are touching, as shown. Carefully and securely tape the cups together with masking tape. For the remainder of this lab, this device that contains the copper is called a *calorimeter*.

6. Find the mass of the stacked cups and the copper. Subtract the mass of the copper to find the mass of the calorimeter. Record this mass in your lab notebook. Push a thermometer into the calorimeter until the bulb is just inside the inner cup. Seal any cracks between the cup and the thermometer with tape.

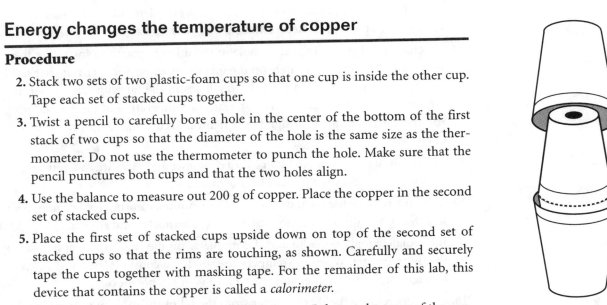

7. Holding the thermometer in place, *slowly* invert the calorimeter so that the shot slides gently down to cover the bulb. Using the appropriate SI units, read and record temperatures at 5.0 s intervals until 5 consecutive readings are the same. Be sure to include all measured digits and one estimated digit.

8. Remove the thermometer from the calorimeter. Push one end of an unsharpened pencil through the thermometer hole until it just blocks the inner hole. Hold it in place with tape.

9. Shake the calorimeter so that the copper falls 425 times from the top of one cup to the bottom of the other cup.

10. Remove the masking tape from the outer cup. Carefully push the thermometer into the calorimeter. Tape any cracks between the cup and the thermometer.

11. Measure the temperature of the copper at the bottom of the calorimeter. Record the temperature in your notebook, using appropriate SI units. Be sure to include all measured digits and one estimated digit.

Analysis

C. How much did the temperature of the copper increase?

D. Use the physics concepts of work, energy, and force to describe what happened to the copper.

E. Were you surprised that the temperature increased? Explain.

Mass and changes in temperature

Procedure

12. Stack a pair of cups one inside the other and tape them together securely. This will make a calorimeter. Place the calorimeter with one lid on a balance and measure its mass. Record the mass in your notebook.

13. Add 200 g of copper shot to the inner cup and determine the mass of the calorimeter and copper. Record the mass in your notebook.

14. Carefully use a pencil to make a hole in the lid big enough to insert the thermometer. Place the lid securely on the inner cup. Insert the thermometer until the bulb touches the copper. Cover any holes in the lid with tape. When the temperature reaches a constant level, read the temperature of the copper using the appropriate number of significant figures. Record the temperature in your lab notebook.

15. Measure 100 mL of hot tap water into a container and measure its temperature. Record the temperature in your lab notebook.

16. Carefully remove the lid and thermometer from the calorimeter, keeping them together. Carefully add the hot water to the calorimeter and copper.

17. Immediately replace the lid and thermometer on the calorimeter and observe the thermometer. When the temperature reaches a constant level, measure the temperature of the copper and record it in your lab notebook.

18. Carefully remove the thermometer from the lid of the calorimeter, leaving the lid in place. Place the calorimeter, water, and copper on a balance and find its mass. Record the mass in your notebook.

HRW material copyrighted under notice appearing earlier in this book.

19. Open the calorimeter, and carefully pour out the water. Place the wet copper in the container provided for this purpose. Dry the calorimeter carefully.

20. Repeat steps 13–19 using dry, unheated copper, and 200 mL of hot tap water.

Analysis

F. Did the temperature of the water increase or decrease?

G. Did the temperature of the copper increase or decrease?

H. How did using different amounts of water affect the temperature change?

Temperature change and phases of matter

Procedure

21. Stack a pair of cups one inside the other and tape them together securely. This will make a calorimeter. Place the calorimeter with one lid on a balance and measure its mass. Record the mass in your notebook.

22. Dry an ice cube with a paper towel and add the ice cube to the inner cup. Determine the mass of the calorimeter and ice. Record the mass.

23. Place the lid securely on the inner cup. Insert the thermometer until the bulb touches the ice. Cover any holes in the lid with tape. When the temperature reaches a constant level, read the temperature of the copper using the appropriate number of significant figures. Record the temperature in your notebook.

24. Measure 100 mL of cold tap water into a container and measure its temperature. Record the temperature in your lab notebook.

25. Carefully remove the lid and thermometer from the calorimeter, keeping them together. Add the cold water to the calorimeter and ice, being careful to avoid spilling any water.

26. Immediately replace the lid and thermometer on the calorimeter and observe the thermometer. When the temperature reaches a constant level, gently shake the calorimeter to make sure the ice is melted. Measure the temperature of the water and record it in your lab notebook. This should take about five minutes.

27. Carefully remove the thermometer from the lid of the calorimeter, leaving the lid in place. Place the calorimeter, water, and ice on a balance and find its mass. Record the mass in your notebook.

28. Open the calorimeter, and carefully pour out the water. Dry the calorimeter.

29. Repeat steps 22–28 using a fresh ice cube and 200 mL of cold tap water.

30. Repeat steps 22–28 using a fresh ice cube and 50 mL of cold tap water.

Analysis

I. As the ice cube melted, did the temperature of the water change?

J. How did using different amounts of water affect the final temperature of the water?

Thermal Conduction

Schlachter Products
Bethel Park, PA

January 24, 2000

Dr. Katherine Loughrey
1% Inspiration Laboratories
14557 West Post Road
Tempe, Arizona 85289

Dear Dr. Loughrey:

It was good to speak with you last week at the Materials Science Conference. I am glad that you too are aware of the current environmental crisis facing our planet. For the past 15 years, my company has been producing personalized thermal products, such as ice chests and thermal stadium pillows. These products are durable, so they are preferable to disposable products that serve the same purpose. Unfortunately, the majority of these products use environmentally unfriendly materials, such as plastic foam. Our goal is to gradually phase out these constituents in favor of other materials without significantly raising our costs.

We are working to develop new environmentally safe polymers that will serve our needs, but we are primarily interested in finding currently available materials that can be used in our product line.

Basically, we are in need of materials that retain heat for long periods of time. We hope you can recommend appropriate materials or inform us of how simple, ecologically sound materials might be modified to retain heat better.

We are also interested in a similar project for a new product. We want to begin producing quick-thaw pans for frozen foods. We need to know of materials that radiate heat quickly. Your recommendations on this matter will be greatly appreciated. If you have any questions, please do not hesitate to send an E-mail or to call.

I hope to speak with you soon. If you have any questions, please do not hesitate to send an E-mail or to call.

Sincerely,

Brian E. Clark

Brian E. Clark

More notes on testing procedures are on page 46.

1% Inspiration Laboratories

MEMORANDUM

Date: January 26, 2000
To: Materials Research Team
From: Katherine Loughrey

Attached is the work request from Schlachter Products. They seem to want both extremes: materials that retain heat for a long period of time and materials that radiate heat quickly. This seems like a fairly basic project, and I think we will be able to do one set of tests to solve both problems. Check out the materials supply list to see what we have available. Also, do some literature searching to find other materials that we might want to get in stock for testing. See me to order some samples; if they're available, test them as well.

Keep in mind that simple materials are less expensive. If modifications such as environmentally sound paint or coating can be made using easily obtained materials, so much the better. Cost is a factor here, so don't seek out exotic new materials. Also, remember that the surface area of the samples will affect the amount of radiated or absorbed energy—standardize your experimental controls. Also be sure to develop a standardized procedure: I have jotted down some ideas on a note card, so make sure you look them over before you prepare your plan.

Before you go into the lab and begin testing, I need to see your plan. Describe the tests you are going to perform in the lab. Include an explanation of how you chose the materials you are going to test.

When your tests are complete, prepare your report in the format of a patent application, describing the tests you performed and analyzing all your results. Your report should give specific recommendations for the materials to be used for the ice chests and also for the quick-thaw pans. Include relevant heating and cooling curves, and include a complete mathematical assessment.

14557 West Post Road • Tempe, Arizona 852

See next page for safety requirements, materials list, and more hints.

MATERIALS

ITEM	QTY.
✔ bulb and socket	1
✔ connecting wires and plug	1
✔ aluminum can	1
✔ black painted metal cup	1
✔ ceramic cup	1
✔ paper cup	1
✔ steel can	1
✔ stopwatch	1
✔ thermometer	2
✔ white painted metal cup	1

SAFETY

- Never put broken glass or ceramics in a regular waste container. Use a dustpan, brush, and heavy gloves to carefully pick up broken pieces and dispose of them in a container specifically provided for this purpose.

- Use a hot mitt to handle resistors, light sources, and other equipment that may be hot. Allow all equipment to cool before storing it.

- If a thermometer breaks, notify the teacher **immediately.**

- Do not heat glassware that is broken, chipped, or cracked. Use tongs or a hot mitt to handle heated glassware and other equipment because it does not always look hot when it is hot. Allow all equipment to cool before storing it.

- If a bulb breaks, notify your teacher immediately. Do not remove broken bulbs from sockets.

The most important thing is to make sure all tests are the same. I think we should use a light source to raise the temperature of each sample. While the sample is exposed to the light, keep track of how its temperature rises. Then remove the light source and measure how the sample's temperature drops. All samples should be the same size and should be placed at the same distance from the light source. Any factors in the lab that could affect one sample differently than others should be eliminated if possible.

HOLT PHYSICS

Discovery Lab

Pendulums and Spring Waves

HRW material copyrighted under notice appearing earlier in this book.

SAFETY

- Attach masses securely. Perform this experiment in a clear area. Falling or dropped masses can cause serious injury.

- Tie back long hair, secure loose clothing, and remove loose jewelry to prevent their being caught in moving or rotating parts.

MATERIALS

✔ 5 metal washers
✔ cord, 1.00 m
✔ long, loosely coiled spring
✔ masking tape
✔ meterstick
✔ paper clip
✔ protractor
✔ stopwatch
✔ support stand and clamp

OBJECTIVES

- Determine the factors that influence the time interval required for a pendulum to complete one full swing.
- Investigate the nature of pendulum and wave motion.

The period of a pendulum

Procedure

1. Construct a pendulum like the one shown at right. Attach a bent paper clip to one end of a 1.00 m cord. Attach the other end of the cord to a clamp that is securely attached to a support stand so that the bottom of the paper clip hangs 0.50 m below the clamp. Securely clamp the support stand to the edge of the tabletop.

2. Hang a small metal washer from the paper clip. Bend the paper clip to hold the washer securely. Remove all obstacles nearby so that the washer is free to swing from side to side.

3. Lift the washer so that the cord is taut between the washer and the clamp. Raise it to a 20° angle from its resting position.

4. Release the washer. Begin the stopwatch the moment the washer is released. Stop timing when the washer completes 10 full swings (over and back). Divide the time by 10 to get the average time interval required for each swing.

5. In your notebook, record the angle, the total time, the number of swings, and the average time required for each swing. Be sure to use the correct number of significant digits and the appropriate SI units.

6. Lift the washer so that the cord is taut between the washer and the clamp. Raise it to a 15° angle from its resting position.

7. Release the washer. Begin the stopwatch the moment the washer releases. Stop timing when the washer completes 10 full swings (over and back). Divide the time by 10 to get the average time interval required for each swing.

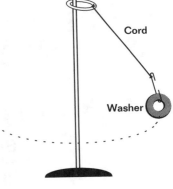

Cord

Washer

8. In your notebook, record the angle, the total time, the number of swings, and the average time for each swing. Be sure to use the correct number of significant digits and the appropriate SI units.

Analysis

A. How much time did it take for the pendulum to complete one full swing when it was raised to a 20° angle?

B. How much time did it take for the pendulum to complete one full swing when it was raised to a 15° angle?

C. Compare the number of seconds of each swing at each position. Which initial angle required the longest time interval to complete one full swing?

The length of a pendulum

Procedure

9. Adjust the cord in the clamp so that the pendulum is longer than 0.50 m but shorter than 1.00 m. Measure and record the length of the pendulum.

10. Lift the washer so that the cord is taut between the washer and the clamp. Raise it to a 20° angle from its resting position.

11. Release the washer. Begin the stopwatch the moment the washer is released. Stop timing when the washer completes 10 full swings. Find the average time interval for one swing.

12. Adjust the cord in the clamp so that the pendulum is longer than 10 cm but shorter than 20 cm. Measure and record the length of the pendulum. Repeat steps 10 and 11.

Analysis

D. How long did it take for each pendulum to complete one full swing?

E. Compare your observations for these pendulums with your observations for the 0.50 m pendulum. Plot your data on a graph of period versus length.

Building a pendulum with a specific period

Procedure

13. Based on your graph and your results above, adjust the length of the pendulum to create a pendulum that requires 1.0 s to complete one full swing.

14. When you have found the correct length, find the time required for a pendulum with the same length and a different mass to complete one full swing. Select two or three washers, and add them to the washer on the paperclip.

15. Measure the time required for the pendulum to complete 10 full swings, and find the average time for one full swing.

Analysis

F. How long was the cord for the pendulum that took 1.0 s to complete one full swing?

G. Did adding mass to the pendulum change the time required for one full swing?

H. To make a pendulum that requires 2.0 s for one full swing, would you lengthen or shorten the cord? Explain your reasoning.

Spring waves

Procedure

16. Hold a long, loosely coiled spring at one end. Have a partner hold the other end of the spring. Place the spring on the floor so that it is straight between both ends.

17. Quickly lift one end of the spring about 30 cm from the floor and place it on the floor again. You should do this in one second or less.

18. Observe the spring. Record your observations in your notebook. Draw a picture in your notebook of what you see. Clearly indicate the direction of motion.

19. Hold the spring at one end. Have a partner hold the other end of the spring. Place the spring on the floor so that it is straight between both ends.

20. Quickly move one end of the spring about 15 cm to the right and then 30 cm to the left. Make sure that the other end remains firmly on the floor.

21. Observe the spring. Record your observations in your notebook. Draw a picture of what you see in your notebook.

22. Hold the spring at one end. Have a partner hold the other end of the spring. Place the spring on the floor so that it is straight between both ends.

23. Quickly push one end of the spring forward and bring it back to its original place.

24. Observe the spring. Record your observations in your notebook. Draw a picture of what you see in your notebook.

Analysis

I. What did you observe when you quickly lifted the spring and set it back down again?

J. What did you observe when you quickly moved one end of the spring about 15 cm to the right and then 30 cm to the left?

K. What did you observe when you quickly pushed the spring forward and brought it back to its original place?

Tensile Strength and Hooke's Law

ORSINO DRUMS

February 3, 2000

Dr. Wes Graham
1% Inspiration Laboratories
14557 West Post Road
Tempe, Arizona 85289

Dear Dr. Graham:

I am writing in regard to my company, Orsino Drums. We are seeking a replacement for the springs used to provide resistance in the foot pedal of the drums we manufacture. The replacements need to be strong and reliable, and the displacement of the spring should be proportional to the force applied. I know that your company has done tensile testing of elastic and non-elastic materials in the past, and I hope that you will be able to provide such a service. I am enclosing a sample of the springs we use so that you can test it to determine what our needs are.

We also sell a low-end practice drum pedal, mostly for beginning drummers. In an effort to keep the prices of these pedals low, we are considering a move toward elastic bands, but we are not sure if their properties make them suitable for a drum-pedal spring. They need to show little sign of fatigue under normal use. I'm enclosing samples of these as well. I am very interested in your thoughts on their utility.

Thanks again for all your help. I look forward to hearing from your company.

Best wishes,

Mike Orsino

Mike Orsino, President

A picture of the
drum pedal is on
page 52.

1% Inspiration Laboratories

MEMORANDUM

February 4, 2000
To: Research and Development Team
From: Wes Graham

This letter is fairly self-explanatory. Test the springs and the rubber bands, and compare their performance. I need to see graphs and values for the spring constant. My hunch is that there is no way that a rubber band will be able to substitute for a spring, but I think that doubling the bands might give a reasonable substitute. Take a look at the marketing information they sent along, with a picture of the drum pedal. You can tell that there are basically two springs (or rubber bands) that provide resistance to the foot of the drummer.

Check this out during the next week, and let me know how the performance of the elastic bands compares with the performance of the springs.

P.S. Make sure you don't damage the spring samples. When the load is removed, the spring should return to its original length. Don't worry about damaging the rubber bands. In fact, you should try to find out how much force they can handle without breaking. Let me know as much as you can about the springs and the rubber bands.

14557 West Post Road • Tempe, Arizona 852

See next page for safety requirements, materials list, and more hints.

continued

MATERIALS

ITEM	QTY.
✔ extension clamp	1
✔ masking tape	1 roll
✔ mass hanger	1
✔ meterstick	1
✔ pad	1
✔ ruler	1
✔ sample elastic bands	2
✔ sample spring	1
✔ set of masses (50 g–1000 g)	2
✔ stopwatch	1
✔ support stand	1

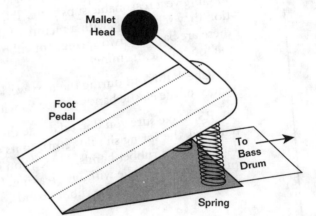

Resonance and the Nature of Sound

SAFETY

- Never put broken glass or ceramics in a regular waste container. Use a dustpan, brush, and heavy gloves to carefully pick up broken pieces and dispose of them in a container specifically provided for this purpose.

- Wear eye protection, and perform this experiment in a clear area. Falling or dropped masses can cause serious injury.

OBJECTIVES

- Explore the phenomenon of resonance in pendulums, and determine what conditions are necessary for resonance to occur.

- Explore the phenomenon of resonance with tuning forks, and determine what conditions are necessary for resonance to occur.

- Discover what variables affect the sound produced by an instrument.

Resonance and pendulums

Procedure

1. Securely suspend two pendulums of different lengths from a flexible rod, as shown. The pendulums should be far enough apart that one pendulum can swing through 20° on each side without touching the other. The longer pendulum should be about 50 cm long. Use a slip knot to attach the pendulum bob to the cord.

2. Raise one pendulum to about a 20° angle so that the cord is taut.

3. Release the pendulum so that it swings freely.

4. Observe both pendulums for one minute as the released pendulum swings. Record your observations in your lab notebook.

5. Adjust the length of the longer pendulum using the slip-knot so that both pendulums are the same length. Make sure that one pendulum can swing without touching the other.

6. Raise one pendulum to approximately a 20° angle so that the cord is taut.

7. Release the pendulum so that it swings freely.

8. Observe both pendulums for one minute as the released pendulum swings. Record your observations in your lab notebook.

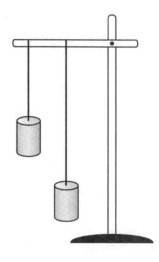

Analysis

A. When both pendulums were different lengths, what happened when one pendulum was raised and released? Describe what happened to the second pendulum.

B. When both pendulums were the same length, what happened when one pendulum was raised and released? Describe what happened to the second pendulum.

C. When both pendulums were swinging, did they have the same frequency or different frequencies? Could you make them swing with different frequencies? Try it and record the results.

Resonance and tuning forks

Procedure

9. Place a tuning fork and resonator box on the table. Select a second resonator box with a tuning fork that is labeled with a different frequency, and place it in line with the first box, as shown. The open mouths of the boxes should be about 50 cm apart.

10. Use a rubber tuning-fork hammer to strike the first tuning fork. Strike the fork swiftly and firmly.

11. Listen to the sound produced by the tuning fork. Listen for any sound produced by the second tuning fork. Record your observations in your lab notebook.

12. Replace one of the tuning forks with another tuning fork that is labeled with the same frequency.

13. Use the rubber tuning fork hammer to strike the first tuning fork.

14. Listen to the sound produced by the tuning fork. Listen for any sound produced by the second tuning fork. Record your observations in your lab notebook.

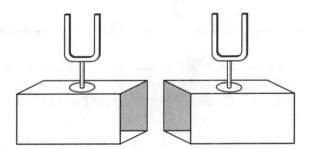

Analysis

D. When the tuning forks had different frequencies, what happened when one was struck? Did you hear any sound produced by the second tuning fork?

E. When the tuning forks had the same frequency, what happened when one was struck? Did you hear any sound produced by the second tuning fork?

Fundamental frequency

Procedure

15. Hold a narrow-mouthed bottle securely. Blow across the top of the bottle to make the bottle produce a whistling sound. Listen to the sound produced.

16. Select a short tube. Wrap masking tape around one end of the tube until the tube will fit snugly in the mouth of the bottle. Do not obstruct the end of the tube. Carefully push the tube firmly into place in the bottle.

17. Blow across the top of the tube in the bottle to cause it to produce a whistling sound. Listen to the sound produced.

18. Carefully remove the tube from the bottle and replace it with a longer tube. Blow across the top of the tube and listen to the sound produced.

19. Carefully remove the tube from the bottle and replace it with a longer tube. Blow across the top of the tube and listen to the sound produced.

20. Remove the tube from the bottle. Carefully pour water into the bottle to a depth of about 2 cm. Blow across the top of the bottle and listen to the sound produced.

21. Add more water to the bottle to a depth of about 4 cm. Blow across the top of the bottle and listen to the sound produced.

22. Continue to add water to the bottle in 2 cm increments until the bottle is full or no longer produces a sound. Listen to the sound produced by blowing across the top of the bottle after each addition.

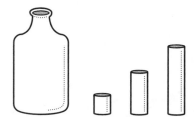

Analysis

F. What happened to the sound as you added tubes of increasing length?

G. What happened to the sound as you added water to the bottle?

H. How did adding tubes affect the total length of the apparatus?

I. How did adding water affect the total length of the apparatus?

J. What is the relationship between the length of the apparatus and the sound produced?

Building a Musical Instrument

EASTSIDE HIGH SCHOOL

February 17, 2000

Ms. Leslie Seecleff
Education Outreach Committee
1% Inspiration Laboratories
14557 West Post Road
Tempe, Arizona 85289

Dear Ms. Seecleff:

Thank you so much for the work you have done organizing the tutors and volunteers in the Education Outreach Committee here in town. Your volunteers have done a lot to help the students keep up with their school work, and I know you have also helped make learning fun!

We are getting ready for our annual Spring Science Fair, which will include students in grades K–12 from all the schools in our district. The volunteer tutors from your labs have always provided a lot of help with the science fair, but this year we have a special project for you. This year our physics classes have all focused on how physics is related to music. Throughout the year, students have attended special presentations about physics and music, including a workshop led by some of your tutors. The theme of the science fair this year is music, and we would like you to help us out by developing some instruments from basic physics principles. We will use these instruments, with reports explaining how physics concepts relate to the design of each instrument, as a special display at the science fair.

Because the focus of the display will be that physics determines how the instruments work, you don't need to worry about using special materials to make them. Simple household items will do. The fair will be held on April 29 in the Eastside High School gymnasium. Thank you so much for your continued support of our program.

Sincerely,

Calvin Saddleback

Calvin Saddleback

More information is on page 58.

1% Inspiration Laboratories

MEMORANDUM

Date: February 22, 2000
To: Education Outreach Committee
From: Leslie Seecleff

It's time for the school district's science fair again, and this year they have asked us to prepare a special exhibit for the students. I think it sounds like a lot of fun. As always, whenever we prepare exhibits for the fair we want to set a good example for the students to follow in their own work. To that end, I have drawn up some guidelines for the instruments.

Each instrument should be homemade and should meet the following requirements:

1. Each volunteer must make one musical instrument.

2. The instrument is to be made from common household materials.

3. The instrument must be capable of producing a complete octave.

4. Each instrument must be accompanied by a patent application that explains the workings of the instrument and describes in detail how physics principles apply to the instrument.

I have gone through the supply room and put together a list of materials that we have available. If you need something else, let me know; we may be able to find it. Before you begin work, please draw up a plan describing what kind of instrument you want to make and how you will use physics to meet the guidelines above.

Also take a look at the flyer for this year's fair. If the flyer is any indication, these students have really made the connection between physics and music this year, so the fair should be exciting.

Good luck and have fun!

14557 West Post Road • Tempe, Arizona 8520

See next page for safety requirements, materials list, and more hints.

MATERIALS

ITEM
- ✔ adhesive tape
- ✔ bottles
- ✔ cans
- ✔ cardboard
- ✔ cord
- ✔ funnel
- ✔ glasses
- ✔ glue
- ✔ pipes of various lengths
- ✔ plastic combs
- ✔ plastic containers
- ✔ pots and pans
- ✔ rubber bands
- ✔ silverware/flatware
- ✔ stones
- ✔ tape
- ✔ wire
- ✔ wood blocks

SAFETY

- Review lab safety guidelines. Always follow correct procedures in the lab.

- Tie back long hair, secure loose clothing, and remove loose jewelry to prevent their getting caught in moving or rotating parts.

Tempe Public Schools

present

The Science of Music

Spring Science Fair 2003

April 29   7:00 P.M.

Eastside High School Main Gymnasium

SEE scientists and engineers

build their own musical instruments.

HEAR the sounds they can produce.

TRY it yourself.

HOLT PHYSICS
Discovery Lab

Light and Mirrors

SAFETY

• Secure all apparatus, and perform this experiment in a clear area. Swinging or dropped masses can cause serious injury.

OBJECTIVES

• Form images using mirrors.
• Locate images using different methods.

Virtual images

Procedure

1. Secure the normal eye chart to the wall using strong tape.

2. Choose any line on the chart, and step back just until the line can no longer be read clearly. Mark the position on the floor where you are standing with masking tape. Label it "reading point."

3. Measure the distance from the eye chart to the reading point with a meterstick. Record this distance in your notebook, using the appropriate SI units. Also record the number of the line that you were trying to read.

4. Secure a small flat mirror against the wall at chest level using strong tape.

5. Place the back of the reverse eye chart against your chest. Position the chart so that the line that you read appears in the mirror. Step back from the mirror, holding the eye chart against your chest until the image of this line is barely readable.

6. Mark the position on the floor where you are standing with masking tape. Label it "new point."

7. Measure the distance from the eye chart to the new point. Record this distance in your notebook, using the appropriate SI units.

Analysis

A. Describe the image of the reverse eye chart you saw on the surface of the mirror. Compare it with the appearance of the normal eye chart.

B. What distance did you measure between the mirror and the reverse eye chart?

C. What distance did you measure between the starting point and the eye chart on the wall?

D. Compare your answers in B and C. What is the relationship between the distances?

MATERIALS

✔ paper
✔ curved mirror
✔ eye charts, both normal and reverse
✔ meterstick
✔ mirror supports
✔ pencil
✔ protractor
✔ ruler or straightedge
✔ small flat mirror
✔ T-pin
✔ tape
✔ white paper

Flat mirrors

Procedure

8. Using two mirror supports, vertically stand one flat mirror on a table, away from the edge, as shown. Place a sheet of white paper on the tabletop so that the front of the mirror faces the paper. Tape the paper and mirror supports to the table so that they do not slide.

9. Using tape, carefully secure a T-pin on the tabletop, with the T side down in front of the mirror. Remove the eraser from a pencil. Secure the eraser on the pin to cover the point.

10. Wearing a pair of safety goggles, move your head to one side of the pin. Close one eye and place your open eye at the level of the tabletop. Observe the image of the pin in the mirror.

11. Use a ruler to draw a straight line on the paper from the image of the pin in the mirror to the position of your eye. Label it "outgoing beam." Use a ruler to draw a straight line from the object to the mirror's surface, connecting with the line labeled "outgoing beam." Label it "incoming beam."

12. Draw a line on the paper from the position of your eye perpendicular to the mirror's surface. Draw a line from the object perpendicular to the mirror's surface. Both lines should be parallel to each other. These lines will form angles with the lines you drew in step 11.

13. Measure the angle between the line labeled "outgoing beam" and the nearest perpendicular line. Measure the angle between the line labeled "incoming beam" and the nearest perpendicular line. Record these angles in your notebook, using the appropriate SI units.

14. Move your eye to a new position. Repeat steps 10–13.

15. Move your eye to a third position. Repeat steps 10–13.

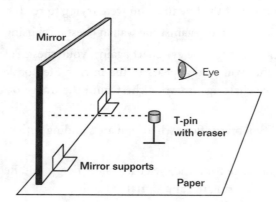

Analysis

E. Compare the two angles measured in step 13 for each position. What is the relationship between the angles?

F. In your notebook, draw the experimental setup as viewed from above. Include the lines and angles for each trial.

Curved mirrors

Procedure

16. Obtain a curved mirror. Use one mirror support to hold the mirror upright on the bench. Place the mirror so that you are facing the side that curves outward.

17 Place an object at various distances from the mirror. Look at the image of the object in the mirror.

18. Observe and record in your notebook how the image appears. Include the object's position (close to the mirror, far from the mirror), the size of the image (enlarged, small), and the orientation of the image (upright, upside down).

19. Turn the mirror around so that you are facing the side that curves inward.

20. Place an object at various distances from the mirror. Look at the image of the object in the mirror.

21. Observe and record in your notebook how the image appears. Include the object's position (close to the mirror, far from the mirror), the size of the image (enlarged, small), and the orientation of the image (upright, upside down).

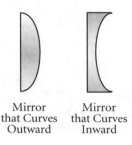

Mirror
that Curves
Outward

Mirror
that Curves
Inward

Analysis

G. How did the image appear when the object was in front of the curved-out mirror?

H. How did the image appear when the object was close to the curved-in mirror?

I. How did the image appear when the object was far away from the curved-in mirror?

Designing a Device to Trace Drawings

Eastern Museum Press

March 29, 2000

Dr. Alexis White
Research and Development
1% Inspiration Laboratories
14557 West Post Road
Tempe, Arizona 85289

Dear Dr. White:

I am in charge of the publishing house here at the Eastern Museum. We publish art books as well as scientific and scholarly books and journals. Recently we have acquired a very old manuscript that is too delicate to be handled or exposed to bright lights.

This manuscript contains many scientific illustrations that we would like to reproduce in a new book. Obviously, this job calls for absolute accuracy. We have come to the conclusion that tracing may be the best method.

We are wondering if you could develop a piece of equipment that causes a virtual image of a picture to appear on a piece of paper next to an artist's real hand so that the artist can trace the image.

I would also greatly appreciate it if you could provide a clear explanation of how the device works so that I can explain its working mechanism to my colleagues. I look forward to hearing from you.

Sincerely,

Caroline Miller

Caroline Miller
Director

A diagram of a related device is on page 64.

1% Inspiration Laboratories

MEMORANDUM

Date: April 1, 2000
To: Optical Design Staff
From: Alexis White

This project seems like one that we can handle. Please start by coming up with a plan for your device. Before you go into the lab, I would like to see a detailed plan including a materials list and a proposed design with ray diagrams.

Caroline's description reminds me of the structure of a periscope, so I suggest looking at the construction of one of these before you begin. I have included a diagram for you to look at while you come up with a plan.

I think you will need to include an eyepiece for the artist to look through during the tracing process. There are some materials on the list that will probably work for the eyepiece in the model. In your final report, include an explanation of why the eyepiece is needed. I would like to know if we could eliminate it and save some money.

In the lab, build a model out of materials that we have readily available. I have included a list of materials that we have on hand for this project. Wear goggles while you work. Your final report should be in the format of a patent application and should include all of the following:

- the model of the tracing device with instructions on how to use it, including information on how far it has to be from the object in order to trace it

- a drawing with objects and images showing how the device works

- an explanation of how the device works

14557 West Post Road • Tempe, Arizona 852

See next page for safety requirements, materials list, and more hints.

MATERIALS

ITEM

✔ adhesive tape
✔ cardboard
✔ converging lens
✔ craft knife
✔ diverging lens
✔ drinking straw
✔ glass
✔ light source
✔ mirror
✔ see-through mirror
  or one-way mirror
✔ support stands and clamps
✔ top from a sports-drink bottle
✔ various hollow cylinders

SAFETY

• Wear eye protection and perform this experiment in a clear area.

• Never put broken glass or ceramics in a regular waste container. Use a dustpan, brush, and heavy gloves to carefully pick up broken pieces and dispose of them in a container specifically provided for this purpose.

• Avoid looking directly at a light source. Looking directly at a light source may cause permanent eye damage.

Periscope

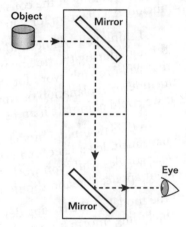

Refraction and Lenses

SAFETY

- Secure all apparatus and perform this experiment in a clear area. Swinging or dropped masses can cause serious injury.

- Avoid looking directly at a light source. Looking directly at a light source may cause permanent eye damage. Always wear eye protection during this exercise.

OBJECTIVES

- Observe how light behaves as it passes from one substance to another.
- Observe images formed by different lenses.

Principles of refraction

Procedure

1. Place a clean sheet of paper on the table. Secure it with tape so that it does not slide.

2. Place a small, clear plastic container on the paper. Carefully trace around the edges of the container, and then remove the container. Draw a line perpendicular to each side of the container. Throughout the lab, you will measure the angle of the incoming beam and the outgoing beam relative to these lines.

3. Carefully pour water into the container until it is half full. Add several drops of milk to the water, and stir carefully. Replace the container on the outline drawn on the paper.

4. Carefully cut a drinking straw so that it is 2.0 cm long. Tape the straw so that it is perpendicular to the flashlight's lens. Cover the rest of the flashlight's face with electrical tape so that light can only exit through the straw when the flashlight is turned on.

5. Use a ruler to draw a line at an angle to one of the perpendicular lines. The line should touch the side of the container.

6. Viewing from above, carefully place the flashlight on the tabletop so that the straw is aligned with the angled line and the beam enters the container. Gently tap two chalkboard erasers together once on each side of the container so that the beam is clearly visible. Observe where the light beam exits the container.

MATERIALS

- ✔ coin
- ✔ drinking straw
- ✔ flashlight
- ✔ medicine dropper
- ✔ milk
- ✔ modeling clay
- ✔ opaque bowl
- ✔ pencil
- ✔ plastic electrical tape
- ✔ protractor
- ✔ ruler
- ✔ small, clear, rectangular container
- ✔ used chalkboard erasers
- ✔ various curved lenses

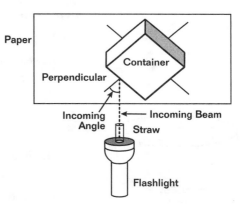

7. Cut a drinking straw so that it is 5.0 cm long. Mount the straw on a small piece of modeling clay so that it is held horizontally at the same height above the table as the flashlight beam. Align this straw on the tabletop with the exiting beam by viewing the beam through the straw.

8. Observe from above the path of the light as it travels through the container. Draw a line on the paper to make the position of the straw. Measure the angle between the beam going into the container (incoming beam) and the perpendicular line with a protractor. Measure the angle between the beam going out of the container (outgoing beam) and the nearest perpendicular line with a protractor. Record your observations and the angles in your notebook.

Analysis

A. Draw the entire setup viewed from above. Include a ray diagram of the light beam and all angles in your drawing.

B. Do the straws lie in a straight line?

C. As the light traveled through the air before it reached the container, did it travel in a straight path or did it bend?

D. As light traveled into the container, did it travel in a straight path or did it bend?

E. As the light traveled through the milky water, did it travel in a straight path or did it bend?

F. As the light traveled from the container to the air, did it travel in a straight path or did it bend?

G. As the light traveled through the air after it left the container, did it travel in a straight path or did it bend?

Seeing around corners

Procedure

9. Place a coin in an empty bowl.

10. Lower your head until the coin goes just out of view. Hold your head in this position while your partner carefully fills the bowl with water without moving the coin.

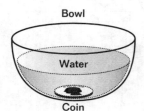

Analysis

H. Draw the view of the setup from the side. Include a ray diagram of the light beam from the coin to your eye in your diagram.

I. As the light traveled through the air before it reached the bowl of water, did it travel in a straight path or did it bend?

J. As light traveled into the bowl of water, did it travel in a straight path or did it bend?

K. As the light traveled through the water, did it travel in a straight path or did it bend?

L. As the light traveled from the bowl to the air, did it travel in a straight path or did it bend?

M. As the light traveled through the air from the bowl of water to your eye, did it travel in a straight path or did it bend?

N. The coin was out of view until the water was added. What do you think happened?

Lenses

Procedure

11. Obtain a lens that is thicker in the middle than at the edges.

12. Place an object at various distances from the lens. Look through the lens at the object.

13. Observe and record in your notebook how the image appears. Include the object's position (close to the lens, far from the lens), the size of the image (enlarged, small), and the orientation of the image (upright, upside down).

14. Obtain a lens that is thinner in the middle than at the edges.

15. Place an object at various distances from the lens. Look through the lens at the object.

16. Observe and record in your notebook how the image appears. Include the object's position (close to the lens, far from the lens), the size of the image (enlarged, small), and the orientation of the image (upright, upside down).

Lens that is thicker in the middle than at the edges.   Lens that is thinner in the middle than at the edges.

Analysis

O. How did the image appear when the object was far away from the lens that is thicker in the middle than at the edges?

P. How did the image appear when the object was close to the lens that is thicker in the middle than at the edges?

Q. How did the image appear when the object was in front of the lens that is thinner in the middle than at the edges?

R. Compare a curved lens with a curved mirror. What similarities and differences are there in the way that light behaves?

Camera Design

DOLLAR A DOZEN PRODUCTS
25300 VILLA LOS LOBOS
ALBUQUERQUE, NEW MEXICO

April 11, 2000

Dr. Lincoln Chun
1% Inspiration Laboratories
14557 West Post Road
Tempe, Arizona 85289

Dear Dr. Chun:

We have just bought the inventory of a bankrupt optical company and are interested in using the lenses in a line of inexpensive cameras. The cameras will contain a single lens, and we will make use of the thin lens equation in their design. The logic of the design is based on the following theory.

For a camera to take quality photographs, it is necessary for a focused image to form on the film. The thin lens equation tells us that $1/p_0 + 1/q_i = 1/f$, where p_0 is the object distance from the lens, q_i is the image distance from the lens, and f is the focal length of the lens. This equation predicts that if an object is far away from the lens, the image will always form one focal length from the lens.

Because of this, we can design very inexpensive cameras if we place the film one focal length from the lens and instruct the photographer to use the camera only for pictures of distant objects. These cameras can be inexpensive because the lens never has to move with respect to the film, so no focusing apparatus is necessary, the photographer just points and shoots.

This is the theory. Our problem is that we do not possess any equipment for testing these lenses. For this reason, we would like you to develop a method to determine the focal length of our lenses. We also need a test apparatus so that we can set a screen one focal length from a lens and find the distance objects must be from the lens to form a focused image on the screen. Our lenses vary in diameter from 38 mm to 50 mm, so please design the apparatus to hold lenses spanning these dimensions.

Sincerely,

Maria Padilla

Maria Padilla

More information about camera design is on page 70.

1% Inspiration Laboratories

MEMORANDUM

Date: April 14, 2000
To: Optical Design Staff
From: Lincoln Chun

I have found some disposable cameras that we can study to help us solve this problem. Take one apart and use the thin lens equation to analyze its design. Determine the focal length of the lens, the distance between the film and the lens, and the minimum distance an object must be from the lens for a focused image to appear on the film. Use an unfrosted light bulb as your object and tracing paper for a screen in this part of your investigation.

Finally, design an apparatus with a lens holder on the front and a screen on the back so that we can check out Dollar-a-Dozen's lenses. The apparatus should allow you to vary the screen's distance from the lens. The lens holder should also accommodate lenses of the sizes mentioned. One potential model would have the user simply place the lens in the holder, point the device at a distant object, and look into the screen on the back to see an image. By changing the distance between the screen and the lens you will be able to determine the focal length of each lens.

Before you go into the lab, prepare a plan describing the apparatus you will use and the tests you will perform in each part of the lab. After I have approved your plan, you can go into the lab and begin testing. When you are finished, prepare a report in the format of a patent application, describing your results.

14557 West Post Road • Tempe, Arizona 852

See next page for safety requirements, materials list, and more hints.

MATERIALS

ITEM	QTY.
✔ black construction paper	1 sheet
✔ black electrical tape	1 roll
✔ black paint (water-based)	1 pint
✔ cardboard box	1
✔ craft knife	1
✔ disposable camera	1
✔ double-sided tape	1 roll
✔ foamboard or mounting board	10 sheets
✔ lens and screen supports	4
✔ magnifier	1
✔ masking tape	1 roll
✔ matte acetate, 3 mL–5 mL	1 sheet
✔ medium-sized paint brush	1
✔ meterstick and supports	2
✔ support stand and clamp	1
✔ tracing paper	1 sheet
✔ unfrosted bulb and socket	1

 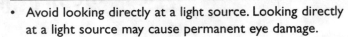
Disposable Cameras: New Old Technology

Since the invention of the camera almost 150 years ago, cameras have come a long way. New technologies over the years have given us cameras with timers, cameras that take the red out of our eyes, and cameras that focus themselves.

Even with all these new developments, however, the basic design of a camera is really the same as it has always been—a camera uses a lens (or other optical device, such as a pinhole) to direct an image onto a light-sensitive material held in a light-proof container.

Because this design is so simple, there is a lot of room for making improvements and developing new features.

Most simple cameras have a fixed focus; this means that there is no way to adjust the focus for objects at different distances from the camera. This kind of camera can focus on most objects, as long as they are at least a certain distance away from the camera. This is the principle behind the new disposable cameras. The newest technology is really old news.

Disposable cameras are very similar to the earliest box cameras. They have a lens that is fixed in focus, and they can be used to take pictures of objects that are a little over 1 m away from the photographer.

After the roll of film has been used, the entire camera is returned to the processor. The film is developed, the pictures are printed, and the camera lens and other parts are recycled and made into new cameras.

Charges and Electrostatics

SAFETY

- Set up all the apparatus securely. Perform this experiment in a clear area. This exercise can produce sparks, so remove flammable liquids from the work area. Carefully handle metal with an insulating material such as rubber gloves to prevent shock.

- Tie back long hair, secure loose clothing, and remove loose jewelry. Roll back long sleeves because they may become charged.

MATERIALS

- ✔ 2 plastic-foam cups
- ✔ balloon
- ✔ comb
- ✔ felt cloth
- ✔ flannel cloth
- ✔ glue
- ✔ large aluminum pan
- ✔ large thick plastic drinking cup
- ✔ meterstick
- ✔ paper clip
- ✔ ruler
- ✔ sheet of aluminum foil
- ✔ silk cloth
- ✔ small aluminum pan

OBJECTIVES

- Discover the electrical properties of metallic and nonmetallic objects.
- Construct an electroscope and investigate how it works.
- Observe forces between charged and uncharged objects.

Constructing an electroscope

Procedure

1. Straighten the broad end of a paper clip. With the paper clip, carefully punch two holes 0.5 cm apart in the center of a small aluminum pan.

2. From a sheet of aluminum foil, cut out two 1 cm × 4 cm strips.

3. With the straightened end of the paper clip, carefully punch a hole 1 cm from one end of each aluminum foil strip.

4. Hook the narrow end of this paper clip through the holes in the strips. The strips should hang parallel to one another as shown.

5. Push the straightened end of the paper clip through a hole in the aluminum pan. Bend the paper clip back so that it can insert into the other hole in the pan. Push the end of the paper clip down through the hole. The paper clip and aluminum strips should hang below the pan.

6. Place a clear, thick plastic cup upright on a tabletop. Set the pan on top of the cup as shown. Throughout this lab, you will observe the movement of the aluminum-foil strips through the cup. This device is referred to as an *electroscope*.

Paper Clip

Aluminum Foil Strips

Plastic Cup

Analysis

A. Did you observe a spark when you touched the aluminum pan?

Determining whether objects are electrically charged

Procedure

7. Inflate a balloon until it measures at least 20 cm from its top to where the knot will be tied. Rinse the balloon with water. Dry it with a paper towel.

8. Vigorously rub the balloon with flannel. Move the balloon toward the electroscope. Closely observe the foil strips hanging from the paper clip. Record your observations in your lab notebook.

9. Touch the part of the balloon that was rubbed with flannel. Move the balloon toward the electroscope, and observe the foil strips. Record your observations.

Analysis

B. What did you observe as the balloon rubbed with flannel moved toward the electroscope?

C. What did you observe as the balloon that you touched moved toward the electroscope?

D. Based on your observations, did a force act on the foil strips? If so, was it a contact force or a field force? Explain.

Observing the effects of electric charge

Procedure

10. Move the aluminum pan toward the electroscope. Observe the foil strips hanging from the paper clip. Record your observations in your lab notebook.

11. Rub the pan with silk. Move the pan toward the electroscope, and observe the foil strips. Record your observations.

12. Glue a plastic-foam cup upside down to the inside of a small aluminum pan as shown. Glue the second plastic-foam cup to the inside of the large aluminum pan.

13. Pick up the small pan by the plastic-foam cup. Rub the pan with the flannel cloth. Move the pan toward the electroscope, and observe the foil strips. Record your observations.

14. Rub an inflated balloon with flannel. Holding only the plastic-foam cup, place the bottom of the small pan firmly against the balloon. Touch the pan once with your finger. Remove the pan from the inflated balloon.

15. Move the pan toward the electroscope, and observe the foil strips. Record your observations.

16. Rub an inflated balloon with flannel. Holding only the plastic-foam cup, place the large pan firmly against the balloon. Touch the large pan once with your finger. Remove the pan from the inflated balloon.

17. Move the large pan toward the electroscope, and observe the foil strips. Record your observations. Place a small ink mark on the edge of the pan's rim.

18. Rub an inflated balloon with flannel. Holding only the plastic-foam cup, place the small pan firmly against the balloon. Touch the small pan once with your finger. This time, place your finger on the plastic-foam cup before removing the small pan from the balloon. Remove the small pan from the inflated balloon.

19. Move the small pan toward the electroscope, and observe the foil strips. Record your observations.

20. Hold both of the pans by the cups. Move the small pan close to the large pan so that the small pan touches only the ink mark on the large pan. Observe what happens as the pans get very close.

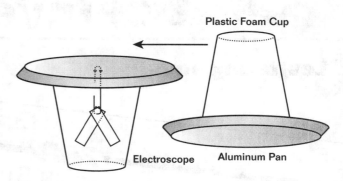

21. Move the large pan horizontally toward the electroscope so that the ink mark is closest to the electroscope. Make sure the pan does not touch any part of the electroscope. Observe the foil strips when the large pan is 1 cm from the device.

22. Rotate the pan 180° so that the side of the pan opposite the ink mark is near the electroscope. Move the pan so that it is 1 cm from the electroscope.

Analysis

E. What did you observe as you first moved the aluminum pan toward the electroscope?

F. What did you observe as the pan rubbed with silk moved toward the electroscope?

G. What did you observe as the pan that you touched moved toward the electroscope?

H. What did you observe as the pan rubbed with flannel moved toward the electroscope?

I. After placing your finger on the plastic-foam cup before removing the small pan from the balloon, what did you observe as the pan moved toward the electroscope?

J. What did you hear as the pans moved close to one another?

K. What did you observe as the ink mark on the pan moved toward the electroscope?

L. After rotating the large aluminum pan so that the ink mark was far from the electroscope, what did you observe as the aluminum pan moved toward the electroscope?

M. Based on your observations, do you think that a force acted on the foil strips?

N. Was this a contact force or a field force?

O. Did the foil strips of the electroscope ever move as you touched the pan?

P. Based on your observations, what was the plastic-foam cup used for?

Levitating Toys

PARAMOUNTAIN STUDIOS
HOLLYWOOD, CA

April 19, 2000

Ms. Colleen Minks
1% Inspiration Laboratories
14557 West Post Road
Tempe, Arizona 85289

Dear Ms. Minks:

We are about to start filming *Hoop Screams,* a science fiction movie featuring small, hovering hoop-shaped structures that attack Earth. We are looking for a company to develop a levitating toy that we can market with the release of the film.

In the movie, the hoops enter other parts of the galaxy by flying through portals constructed out of material excavated from white dwarfs. The hoops can fly back and forth through the portals.

With this background in mind, we would like you to try to develop a toy with a levitating hoop that can be maneuvered around the room. The toy should also include large portals that the hoop can fly through.

If you feel that your company can supply us with this product, please contact me immediately. Thank you for your time. I look forward to hearing from you.

Sincerely,

Monali Jhaveri

Monali Jhaveri

Ideas for the design of the toy are on page 76.

1% Inspiration Laboratories

MEMORANDUM

Date: April 20, 2000
To: Research and Development
From: Colleen Minks

I like the concept for this toy immensely. I think that you should be able to design and build this toy to use electrostatic repulsion.

The flying parts of the toy will have to be made from a lightweight material. We have a few space blankets in our storeroom that I think will be ideal. I have done a little research, and I think an electrophorus can be used to keep the pieces hovering. I have included a diagram of a simple model. The flying hoop and the electrophorus will have to have the same charge for this to work.

Start by constructing the levitating hoop. Construct it as neatly and symmetrically as possible. If you can get this small hoop to levitate, construct a portal for the hoop to hover through. For the best distribution of charge, the portal should not have any sharp edges. Determine whether the portal should be made from an insulator or a conductor, or if it should be grounded. To insulate the portal from its surroundings, hang it from some polyester thread.

If the hoop and portal work, spend some time maneuvering the hoop around the room and through both sides of the portal. Use your observations to make improvements to the design.

Before you go into the lab, give me a plan describing the procedure you will follow in the lab. Your plan should include the method you will use for charging the different parts of the toy.

When you are finished, I would like you to fill out a patent application for the finished toy. The report should include a short explanation of the electrostatic principles involved in the experiment and how they lead to the toy's failure or success. Include drawings showing the distribution of charges on the items.

14557 West Post Road • Tempe, Arizona 852

See next page for safety requirements, materials list, and more hints.

MATERIALS

ITEM	QTY.
✔ plastic-foam plate or polystyrene board	1
✔ new unsharpened pencil with eraser end	1
✔ plastic-foam cup	1
✔ aluminum pie pan	1
✔ wool flannel	1 piece
✔ space blanket	1 small piece
✔ coat hanger	1
✔ polyester thread	1 m
✔ rubber cement	1 jar
✔ meterstick	1
✔ craft knife	1
✔ cardboard or poster board	1 piece

Electrophorus

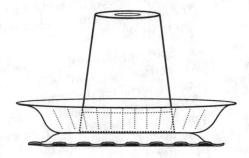

The electrophorus shown in this diagram is constructed from a plastic-foam dinner plate, an aluminum pie pan, and an insulating handle. The handle may be made by gluing a plastic-foam cup upside down on the inside of the pie pan. Charge the plastic-foam plate by rubbing it with a wool cloth. Place the pie pan onto the charged foam, then touch the pie pan with your finger. The pan is now charged and can be used for electrostatic experiments.

Discovery Lab

Resistors and Current

SAFETY

Never close a circuit until it has been approved by your teacher. Never rewire or adjust any element of a closed circuit. Never work with electricity near water; be certain that the floor and all work surfaces are dry.

If the pointer on any kind of meter moves off scale, open the circuit immediately by opening the switch.

Do not attempt this exercise with any batteries or electrical devices other than those provided by your teacher for this purpose.

MATERIALS

- ✔ 6 battery holders for D-cells
- ✔ 6 D-cell batteries
- ✔ light bulb and socket
- ✔ 2 multimeters or 1 dc ammeter, 1 ohmmeter, and 1 voltmeter
- ✔ graph paper
- ✔ insulated connecting wires
- ✔ 2 resistors
- ✔ switch

OBJECTIVES

- Measure current, resistance, and potential difference across various resistors.
- Graph the relationship between the potential difference and current for various resistors.
- Interpret graphs relating potential difference and current for various resistors.

Potential difference and current in a resistor

Procedure

1. Use the multimeter or ohmmeter to measure the resistance of the resistor. Hold the plastic part of the resistance meter probes. Place the probes across the terminals of each resistor in turn by touching the metal part of the red probe to one terminal and the metal part of the black probe to the other terminal. Read the values for the resistance across the resistors, and record the values in your lab notebook using the appropriate SI units.

2. Using the resistor with the larger value, set up the apparatus as shown. Use one wire on each end of the battery holder to connect the switch and the resistor as shown. Connect the switch to one post of the current meter. Connect the other post of the current meter to the resistor. Carefully connect one post of the voltage meter to one side of the resistor, and connect the other post of the voltage meter to the other side of the resistor.

3. Carefully place a battery in the battery holder. **Do not close the switch until your teacher approves your circuit.**

4. When your teacher has approved your circuit, close the switch. Read the value for the current in the resistor, and record the value in your lab notebook using the appropriate SI units.

Battery in
Battery Holder

Current
Meter

Switch

Resistor

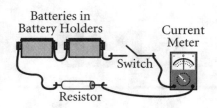

Batteries in Battery Holders

Current Meter

Switch

Resistor

5. Use the multimeter or voltage meter to measure the potential difference across the resistor, and record the value in your lab notebook using the appropriate SI units. Open the switch.

6. Disconnect one side of the battery holder, and add another battery holder to the setup using another wire as shown. Carefully place a battery in the empty battery holder. ***Do not close the switch until your teacher approves your circuit.***

7. When your teacher has approved your circuit, close the switch. Measure the current and the potential difference as in steps 4–5.

8. Disconnect one side of the battery holder, and add a third battery in a row with the other two as in step 6.

9. When your teacher has approved the circuit, close the switch. Measure the current and the potential difference as in steps 4–5.

10. Disconnect one side of the battery holder, and add a fourth battery in a row with the other three as in step 6.

11. When your teacher has approved the circuit, close the switch. Measure the current and the potential difference as in steps 4–5.

12. Replace the resistor with the second resistor, and repeat steps 2–11.

Analysis

A. How did adding more batteries to the setup affect the potential difference measured across the resistor?

B. How did adding more batteries to the setup affect the current measured in the resistor?

C. For each data set, divide the potential difference by the corresponding current. Record these ratios in your lab notebook.

D. Compare the value of the ratios for each resistor. What do you notice?

E. For each resistor, make a graph current on the *x*-axis and potential difference on the *y*-axis. Label each axis with the appropriate SI units.

F. Determine the slope of each graph. To do this, choose one point at the beginning of the graph and one point at the end. Find the change in potential difference and the change in current between these two points. Divide the difference in potential difference by the difference in current.

G. Compare the value for the slope with the resistance that you measured. Are there any similarities?

H. Compare the value for the slope with the ratios found. Are there any similarities?

Potential difference and current in a light bulb

Procedure

13. Place a light bulb securely in a socket. Holding the plastic part of the probes, carefully touch the resistance meter probes across the posts of the socket. Read the value for the resistance across the light bulb, and record the value in your lab notebook using the appropriate SI units.

14. Set up the apparatus as shown. Connect the switch, the battery holder, and the light bulb as shown. Connect one post of the current meter to the switch. Connect the other post of the current meter to one post of the light bulb socket. Carefully connect one post of the voltage meter to one post of the light bulb socket, and connect the other post of the voltage meter to the other post of the light bulb socket.

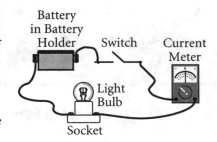

15. Carefully place three batteries in the empty battery holder. ***Do not close the switch until your teacher approves your circuit.***

16. When your teacher has approved your circuit, close the switch. Measure the current in the light bulb, and record the value in your lab notebook using the appropriate SI units.

17. Use the voltage meter to measure the potential difference across the light bulb, and record the value in your lab notebook.

18. Disconnect one side of the battery holder, and add another battery holder to the setup. Carefully place a battery in the empty battery holder. ***Do not close the switch until your teacher approves your circuit.***

19. Measure the current and the potential difference as in steps 16–17.

20. Add a fifth battery in a row with the other two as in step 18.

21. Measure the current and the potential difference as in steps 16–17.

Analysis

I. How did adding more batteries to the setup affect the potential difference measured across the light bulb?

J. How did adding more batteries to the setup affect the current measured in the light bulb?

K. For each data set, divide the potential difference by the corresponding current. Record these ratios in your lab notebook.

L. Compare the value of the ratios. What do you notice?

M. Graph your data with current on the *x*-axis and potential difference on the *y*-axis. Label each axis with the appropriate SI units.

N. Compare the graph with the graphs in E. Are the shapes similar?

O. Determine the slope of the graph. To do this, choose one point at the beginning of the graph and one point at the end. Find the change in potential difference and the change in current between these two points. Divide the change in potential difference by the change in current for the same interval.

P. Compare the ratios with the resistance that you measured. Are there any similarities?

Q. Compare the value of the slope with the ratios found. Are there any similarities?

Battery-Operated Portable Heater

Leaping Lizards
2378 Whippoorwill Road
Bethel, Maine 04217

April 16, 2000

Dr. Ryan Williams
Research and Development
1% Inspiration Laboratories
14557 West Post Road
Tempe, Arizona 85289

Dear Dr. Williams:

My pet store specializes in reptiles and amphibians. In the store, all our cages and aquariums are equipped with hot rocks, heating pads, and basking lights. We need a way to keep the animals warm when they are moved from cage to cage and when they are taken home by their new owners. We are looking for a company to design a small battery-powered cage heater for our lizards and other coldblooded animals. These heaters will be used to heat the animals' carrying cases during cool weather. They can also be used as backups in case we lose power during a winter storm.

I have enclosed some information that we give to new owners, describing how to care for lizards. This information provides the ideal temperature range and some ideas for providing heat and light in the animals' cages. The cage heaters should maintain temperatures somewhere within the temperature range described in the information.

Thank you very much for your attention to this matter. These heaters will solve a serious problem for us and our customers.

Sincerely,

Terry Murphy

Terry Murphy

Information on lizard cages is on page 82.

1% INSPIRATION LABORATORIES

MEMORANDUM

Date: April 18, 2000
To: Research and Development
From: Ryan Williams

I think that we should be able to design these little heaters. The simplest approach is probably to use batteries to provide current in a high-resistance wire placed in the lizard's cage. You should start by controlling the size and number of batteries—use two new D-cell batteries to power the circuit. This will allow you to focus your experiments on determining the length and type of wire necessary to achieve an appropriate temperature. Terry Murphy sent some information on the best temperature range for most of their animals, so refer to that while you work. Let's assume that cool weather is around 10° C and design the heater to raise the temperature 10° C–15° C above the ambient temperature.

Before you begin work, I will need to see your plans for building and testing the device. Make sure you include plans to measure the temperature level of the wire when there is current. Describe how you will raise and lower the temperature to bring it to the appropriate level. Once you discover how to obtain a suitable temperature, you will need to develop a plan for using the wire and batteries safely and effectively in a small animal carrier.

Because raising the temperature of the wire requires a lot of current, you should keep track of the number of times that the batteries are used and approximately how hot the wire gets on each trial. Batteries can be used up very rapidly when they are used to bring wire to high temperatures, so keep careful track of this while you work.

When you are finished, submit your final design to me in the form of a patent application. Your application should include a discussion of the physics principles that describe how the heater works, as well as some explanation of how different types of wire could be used to make similar heaters.

14557 West Post Road • Tempe, Arizona 852

See next page for safety requirements, materials list, and more hints.

MATERIALS

ITEM	QTY.
✔ Nichrome wire	2.5 m
✔ insulated connecting wire	1.5 m
✔ battery holder for 2 D-cell batteries	1
✔ D-cell batteries	2
✔ felt, 20 cm × 60 cm	1
✔ masking tape	
✔ connectors for wire	
✔ multimeter or dc ammeter with connecting leads	1
✔ thermometer or CBL system with temperature probe	1
✔ liquid crystal thermometer strip	1 (opt)
✔ cardboard box	1
✔ bare copper wire	1 m
✔ wire leads with alligator clips	2
✔ insulating materials	
✔ stopwatch	1

SAFETY

Wire coils may heat up rapidly during this experiment. If heating occurs, open the circuit immediately and handle the equipment with a hot mitt. Allow all equipment to cool before storing it.

Never close a circuit until it has been approved by your teacher. Never rewire or adjust any element of a closed circuit. Never work with electricity near water; be sure the floor and all work surfaces are dry.

If the pointer on any kind of meter moves off scale, open the circuit immediately.

Do not attempt this exercise with any batteries or electrical devices other than those provided by your teacher for this purpose.

Caring for your pet lizard

Now that you have your lizard, there are a few things you need to know to make sure your pet lives a happy and healthy life. You are responsible for meeting the dietary, temperature, and habitat needs of your pet. Some lizards, such as iguanas, can live to be over thirty years old, so this is a serious commitment!

Your lizard needs a terrarium that contains places to climb and hide, a water bowl that is easy to get in and out of, and a heater or basking light. A branch or shelf placed directly below the basking light will allow the lizard to quickly raise its body temperature. Lizards also need regular exposure to sunlight or UVB lighting.

Most lizards are active during the day and rest during the night. The ideal temperature range is from 18°C –24°C at night and 21°C –26°C during the day, with a basking area of 29°C –32°C.

HOLT PHYSICS
Discovery Lab

Exploring Circuit Elements

SAFETY

Never close a circuit until it has been approved by your teacher. Never rewire or adjust any element of a closed circuit. Never work with electricity near water; be certain the floor and all work surfaces are dry.

Do not attempt this exercise with any batteries or electrical devices other than those provided by your teacher for this purpose.

If a bulb breaks, notify your teacher immediately. Do not remove broken bulbs from sockets.

MATERIALS

- ✔ 1.5 V flashlight batteries, 2 or 3
- ✔ 5 miniature light bulbs
- ✔ 5 miniature light sockets
- ✔ 20 connecting wires
- ✔ capacitor
- ✔ rubber bands or tape

OBJECTIVES

- Construct circuits using different combinations of bulbs, batteries, and wires.
- Observe the effects of an electric current.
- Compare your observations from different trials to discover how relationships are affected by changing one or more variables.
- Classify and analyze your observations.

Simple circuit

Procedure

1. Place the bulbs securely in the sockets. Using one light bulb, a battery, and wires, connect the bulb to the battery to produce light.

2. Observe how brightly the bulb is lit. Also make observations of other qualities—temperature, sound, smell, color, motion, and anything else you observe. Hold your finger against the insulated part of the wire to test for motion in the wire.

3. Disconnect the battery. In your lab notebook, write a brief description of the bulb's brightness and of your other observations.

Analysis

A. How would you describe the brightness of the bulb?

B. Develop a system for comparing the brightness of different bulbs. Explain how your system would work in different situations, such as in a dark room and in direct sunlight.

C. Other than light, what effects did you observe when the bulb was lit?

D. Based on your observations, how can you detect the presence of current?

Circuit with bulbs in series

Procedure

4. Connect all three sockets of bulbs in a side-to-side row using two wires as shown. Use an additional wire on each end to connect the unattached ends to the battery so that all the bulbs light. (Hint: You may need to use more than one battery to get the bulbs to light.)

5. Compare the bulbs in terms of brightness and other qualities, and then compare the bulbs with the single bulb you observed earlier.

6. Disconnect the battery to sketch your circuit, and briefly describe your observations and comparisons in your lab notebook.

7. Reconnect the battery to light all three bulbs. Unscrew one bulb, and observe the effects on the other bulbs. Try this with the other bulbs to see if the position of the bulb makes a difference.

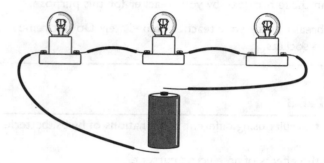

Analysis

E. When all three bulbs were lit, how did the brightness of the bulbs compare? How did the brightness of the bulbs compare with the brightness of the one-bulb system you observed before?

F. What happened when you unscrewed one of the bulbs? Did it matter which bulb you removed? Explain why or why not.

G. Based on your observations, what do you think would happen to the brightness of the bulbs if you added two more bulbs? Explain your reasoning. If time permits, get your teacher's permission and try it.

H. If the brightness of each bulb depends on the current, what do your observations tell you about the current in each bulb in the three-bulb circuit? Is the current the same in each bulb? Why or why not?

I. Suppose that a light bulb provides resistance to the current. How does using more than one light bulb affect the total resistance of the entire circuit? How does it affect the total current?

Circuits in parallel branches

Procedure

8. Connect all three sockets of bulbs in a column, with two wires connecting each pair of sockets, as shown. Each post will be connected to two wires. Using two more wires, connect the posts of the socket to the end of the battery so that all three bulbs light.

HRW material copyrighted under notice appearing earlier in this book.

9. Observe the brightness and other qualities of the bulbs. Compare the bulbs with each other, and with the bulbs you have already observed.

10. Disconnect the battery and record your observations. Draw your circuit in your notebook.

11. Reconnect the battery so that all three bulbs relight. Unscrew one bulb and observe the effects on the other bulbs. Try this with the other bulbs to see if the position of the bulb in the circuit makes a difference.

12. Disconnect the battery and record your observations.

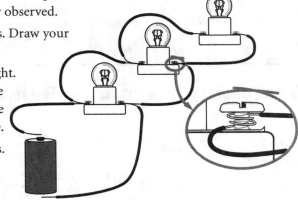

Analysis

J. When all three bulbs were lit, how did the brightness of the bulbs compare? How did the brightness compare with that of other systems you have observed?

K. What happened when you unscrewed one of the bulbs? Did it matter which bulb you removed? Explain why or why not.

L. Based on your observations, what do you think would happen to the brightness of the bulbs if you added two more branches with bulbs? What if the bulbs were different wattages from the bulbs you are already using?

M. If the brightness of a bulb depends on the current, what do your observations tell you about the current in each bulb? Is the current the same in each bulb? Why or why not?

N. Suppose that a light bulb provides resistance to the current. How does using a different branch for each bulb affect the total resistance of the entire circuit? How does it affect the total current?

Circuit with a capacitor

Procedure

13. Your teacher will supply you with a capacitor. Connect the capacitor and a light bulb with wires as shown.

14. Connect the battery to the bulb and capacitor so that the bulb lights. Leave the battery connected until the light goes out. Record your observations.

15. Next, remove the battery and connect the ends of the wires to each other. Observe what happens. Record your observations in your notebook.

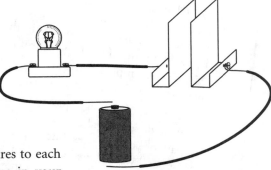

Analysis

O. Describe your observations of the brightness of the bulb when the bulb and capacitor were connected to the battery. Give an explanation of what happened.

P. What happened when the battery was removed and the wires were connected to each other? Explain.

Q. Based on your observations, do you think the current remained constant in this circuit? Explain your answer.

R. What do you think happened to the current when you removed the battery and reconnected the wires? Explain.

Designing a Dimmer Switch

Greenville Historical Science Foundation
Greenville, North Carolina

April 21, 2000

Dr. Kelly Maxwell
1% Inspiration Laboratories
14557 West Post Road
Tempe, Arizona 85289

Dear Dr. Maxwell:

We are restoring the home of Seelie Charles, an inventor who lived in our town during the turn of the century. We are planning to open the house as a historical museum. To promote interest in Ms. Charles's inventions and to give an idea of her vision of the future, we are incorporating many of her inventions and patents into the design of the house.

During the renovation of the laboratory, we came across the enclosed page of Ms. Charles's notebook. After conducting a patent search, we came to the conclusion that Ms. Charles was never able to develop her ideas for a dimming light switch. We would like to use these switches in the house. Electricity will be supplied to the house by means of a DC generator and rechargeable batteries, much like the ones Ms. Charles implemented in 1886 when she became the first citizen in Greenville to use electric lighting in her house.

Because of the high quality of your work, we would like you to develop a lighting design based on Ms. Charles's ideas. Unfortunately, the Foundation is unable to pay for your services, but you will maintain ownership of any patent issued on your design and we will credit you with a plaque at the house, as well as a formal mention in all of our advertising and promotional materials.

Please let me know as soon as possible if you will be able to complete this work on these dimmer-switch lighting systems. Thank you very much for your time.

Sincerely,

K. Azielinski

K. Azielinski

The page from Ms. Charles's notebook is on page 88.

1% INSPIRATION LABORATORIES

MEMORANDUM

Date: April 24, 2000
To: Development Team
From: Kelly Maxwell

I am a real history buff, and this looks too good to pass up. The publicity will be great, and if we get a patent, that could turn out to be profitable too. Ms. Charles has provided us with several good hints, but there are still several pieces to fit together before we start testing. Before you go into the lab, prepare a plan for each of the three designs mentioned in Ms. Charles's notebook:

- a light that can shine at three different brightness levels, with the amount of current controlled by the potential difference supplied

- a light that can shine at three different brightness levels, with the amount of current controlled by the amount of resistance in the circuit

- a light that stays on for a short amount of time, gradually growing dimmer until it is completely dark

I will approve your plan before you start work in the lab, so get this to me as soon as possible. For each design, your plan should include a list of materials needed, a diagram, and a one- or two-sentence explanation of what you expect to happen. I have included a list of the electrical components and equipment we have available. If you need something that you can't find on the list, be sure to ask about it; there may be more equipment available.

You will prepare your report in the form of a patent application. Remember to document all your testing and development procedures in your lab notebook. Good luck!

14557 West Post Road • Tempe, Arizona 852

See next page for safety requirements, materials list, and more hints.

MATERIALS

ITEM	QTY.
✔ metal paper clips	1 box
✔ rubber bands	1 box
✔ tape	1 roll
✔ 1.5 V flashlight battery and battery holder	3
✔ 6.0 V lantern battery	1
✔ capacitor—1 F	1
✔ resistor—390 kΩ	3
✔ resistor—180 kΩ	3
✔ resistor—10 Ω	2
✔ miniature light bulbs, 1.5 V	2
✔ miniature light sockets	2
✔ miniature light bulbs, 2.5 V	2
✔ miniature light bulbs, 6.3 V	2
✔ connecting wires with alligator clips	20
✔ single-throw knife switch	3
✔ double-throw knife switch	2

SAFETY

Never close a circuit until it has been approved by your teacher.

Never rewire or adjust any element of a closed circuit.

Never work with electricity near water—be sure the floor and all work surfaces are dry.

Do not attempt this exercise with any batteries or electrical devices other than those provided by your teacher for this purpose.

May 19, 1901

Idea 1: Design lights that are not always the same brightness, but can be made brighter or dimmer depending upon mood or activity.

Principle: The brightness of a light depends on the amount of current in the circuit, which depends on voltage and resistance.

Design: A circuit could be constructed that contains a different amount of current depending upon the position of a switch. This can be constructed by providing different resistance or different potential difference.

Idea 2: Design a light that will "turn off by itself"—this will be perfect for use in bedrooms, corridors, etc., wherever it is now necessary to walk across a darkened room or use a candle.

Principle: The duration of a light depends on the current in the circuit, which depends on the voltage.

Design: A circuit could be constructed in which the potential difference gradually becomes equal to zero; the current therefore decreases and the light gradually goes out.

Magnetism

SAFETY

Perform this experiment in a clear area. Falling or dropped masses can cause serious injury.

Do not attempt this exercise with any batteries, electrical devices, or magnets other than those provided by your teacher for this purpose.

Never place fingers between the poles of magnets.

MATERIALS

- ✔ 2 bar magnets with labeled poles
- ✔ 4 large, rectangular erasers
- ✔ aluminum foil strips
- ✔ cardboard, 1 sheet
- ✔ compass, 1.8 cm diameter
- ✔ graph paper
- ✔ iron filings in a shaker
- ✔ paper
- ✔ paper clips
- ✔ plastic pen
- ✔ plastic cup
- ✔ rubber band
- ✔ metric ruler
- ✔ scissors
- ✔ staples

OBJECTIVES

- Investigate the properties of the field surrounding a magnet.
- Relate distance and field strength of a magnet.

The nature of magnets

Procedure

1. In each hand, hold a bar magnet by its center. Point the ends labeled *N* toward each other. Slowly move them close together, but do not let them touch. Observe what happens, and record your observations in your notebook.

2. Point the ends labeled *S* toward each other. Slowly move them close together, but do not let them touch. Observe what happens, and record your observations in your notebook.

3. Still holding the magnet by its center, rotate one magnet 180° and point the N end toward the S end of the other magnet. Slowly move the magnets close together, but do not let them touch. Observe what happens and record your observations in your notebook.

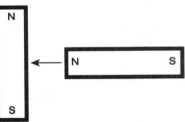

4. Rotate one magnet 90° so that the magnets are perpendicular to one another, as shown. Point the N end of one magnet toward the center of the other magnet. Observe what happens, and record your observations.

5. Repeat step 4, pointing the N end of one magnet toward several different points along the side of the other magnet. Observe what happens, and record your observations in your notebook.

6. Repeat steps 4–5, pointing the S end of one magnet toward the center of the other magnet. Observe what happens, and record your observations in your notebook.

7. Move one end of the magnet toward a paper clip. Observe what happens, and record your observations in your notebook.

8. Repeat step 7 with a variety of different objects, including a plastic cup, a pen, staples, aluminum foil, a rubber band, a pair of scissors, and paper.

Analysis

A. Do the magnets have to be in contact to interact?

B. Was the interaction stronger between the magnets when they were perpendicular or when the magnets were held with their ends facing one another?

C. Which ends of the magnet repelled one another?

D. Which ends of the magnets attracted one another?

E. Classify the objects that were attracted and repelled by the magnet. What did the objects attracted to the magnet have in common?

Mapping a magnetic field

Procedure

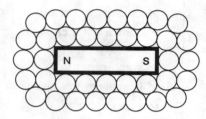

9. Place a sheet of paper on a nonmetallic tabletop. Place a bar magnet in the center of the paper so that the N end of the magnet points to the right. Make sure that the magnet is far away from any other magnets. Trace around the magnet and label the ends *N* and *S* on the paper.

10. Place a small compass on the paper beside the magnet. Trace a circle around the compass with a pencil.

11. Move the compass to a new position beside the magnet, and repeat step 10. Continue until you have traced a pattern of circles around the magnet as shown.

12. Move the compass far away from the magnet. Observe which way the needle points.

13. Place the compass in one of the circles on the paper. Mark the edge of the circle to indicate the direction that the needle points. Remove the compass. Draw an arrow in the circle to represent the position of the compass needle. The tip of the arrow should touch the mark on the edge of the circle, and the tail of the arrow should pass through the center of the circle.

14. Repeat step 13 until all the circles contain arrows.

Analysis

F. Does the compass needle always point the same direction?

G. Does the compass needle always point to the same end of the magnet?

H. Which end of the bar magnet does the compass needle point toward? Which end of the bar magnet does the needle point away from?

I. What kind of force causes the compass needle to change direction, a contact force or a field force?

The shape of a magnetic field

Procedure

15. Place the bar magnet on a nonmetallic tabletop. Make sure that the magnet is far away from any other magnets. Place a sheet of cardboard on top of the bar magnet so that the magnet is under the middle of the cardboard. Support the cardboard at the edges with rectangular erasers so that it remains level. Place one sheet of paper on top of the cardboard.

16. Carefully sprinkle iron filings on top of the paper over and around the magnet.

17. Carefully tap the cardboard a few times. When the filings settle into position, observe the pattern formed.

18. Draw the pattern of iron filings in your lab notebook.

Analysis

J. Compare the pattern made by the iron filings with the pattern of the arrows made by the compass needle. Does the iron-filing pattern have any relationship to the pattern of the arrows?

K. Did it require a force to move the iron filings into position? If so, was it a contact force or a field force?

The strength of a magnetic force

Procedure

19. Place a sheet of graph paper on a nonmetallic tabletop. Place a bar magnet in the center of the graph paper. Make sure that the magnet is far away from any other magnets.

20. On the graph paper, mark positions next to the magnet, as shown. Label these positions A–G.

21. Move a compass to each position on the graph paper. Observe how quickly the compass needle moves at each position. Using the words *strong, medium,* and *weak,* label how the force that moves the compass needle varies at each position.

22. On the graph paper, measure and mark a distance of 3.0 cm from each marked position, as shown.

23. Place a paper clip on a position marked 3.0 cm from the magnet. Point the end of the paper clip toward the magnet.

24. Slowly move the magnet toward the paper clip until the paper clip begins to move toward the magnet. Mark the position of the magnet on the paper. Using appropriate SI units, measure the distance the magnet was from the paper clip. Record the measurements in your lab notebook.

25. Repeat steps 23–24 for each position marked 3.0 cm from the magnet.

Analysis

L. Is the strength of the force the same everywhere, or does it vary along the length of the magnet? Explain.

M. Is the force that caused the paper clip to start moving a contact force or a field force?

Designing a Magnetic Spring

AQUACHEX ENVIRONMENTAL SERVICES

2240 ARENA DRIVE

EVERGREEN, CO 80436

May 20, 2000

Dr. Belinda Fu
Product Development
1% Inspiration Laboratories
14557 West Post Road
Tempe, Arizona 85289

Dear Dr. Fu,

Aquachex Environmental Services specializes in monitoring lakes, streams, and other bodies of water. We have developed a sampling probe that remains in the water for 24 hours at a fixed depth and absorbs certain pollutants. The probe is then recovered and replaced. The used probe is brought to our laboratory for analysis.

Our problem is that the probe frequently hits the bottom too forcefully. This damages the probe and causes the samples to be contaminated with mud. Using a spring at the bottom of the line to slow the probe has failed because the spring corrodes after a few months in the water, especially in more-polluted locations.

We hope that you can develop a magnetic device that will act like a spring to slow the probe as it approaches the bottom and hold it about 20 cm above the bottom.

We are also having difficulty recovering the probe after the 24-hour testing period because the line we use to pull the probe to the surface often becomes tangled with the anchor line or with weeds. We would like your design to include something we can lower to retrieve the probe, so that the line does not have to be left in place during the testing period.

We would like to solve these problems as quickly as possible. We look forward to seeing your design soon.

Sincerely,

Cecil Dawkins

Cecil Dawkins

A page from the Aquachex Field Manual is on page 94.

1% INSPIRATION LABORATORIES

MEMORANDUM

Date: May 25, 2000
To: Product Development Team
From: Belinda Fu

I know we have a good supply of ring magnets in stock. I think we can solve both the spring problem and the retrieval problem using magnets; we can use repulsion between magnets for the spring and attraction between magnets for a hook to grab the probe during retrieval. Write a plan for your design and testing procedures. I will approve your plan before you go into the lab.

In the lab make a model of the probe described in the *Aquachex Field Manual*, and use clay to simulate the resin packages. The finished probe should have a mass of 120 g, just like the real one.

The probe is designed for water 0.5 m to 2.0 m deep, so test your spring design by having the probe slide down a vertical anchor line for 2.0 m in the lab. Remember that the probe will fall faster in air than in water, so any system that works well in air will have an added safety factor when used in water. Test your retrieval system on the same anchor line.

Try several combinations of magnets to optimize both spring force and support height. To keep the cost of the final device down, don't use more than eight ring magnets in your design. If you decide to attach magnets to the probe, remember to make them easily removable so they can be reused when the rest of the used probe is disposed of.

When you are finished, submit your report in the form of a patent application.

14557 West Post Road • Tempe, Arizona 852

See next page for safety requirements, materials list, and more hints.

MATERIALS

ITEM	QTY.
✔ bar magnets	6
✔ ceramic-ring magnets	8
✔ clay	150 g
✔ cord	5 m
✔ craft knife	1
✔ glue	1 tube
✔ heavy plastic cup, 14 oz	1
✔ lids for 12 oz plastic-foam cups	2
✔ meterstick	1
✔ plastic drinking straw	1
✔ plastic water pipe	15 cm
✔ plastic-foam hot drink cup, 12 oz	1
✔ rare earth magnet	1
✔ self-adhesive plastic tape	1 roll
✔ slotted masses, 10 g–500 g	1 set
✔ steel screw	2
✔ waterproof tape	1 roll

SAFETY

- Attach masses securely. Perform this experiment in a clear area. Swinging or dropped masses can cause serious injury.

- Magnets can generate strong forces. Never place your fingers between two magnets.

Aquachex Field Manual

An inexpensive probe using a plastic-foam hot-drink container has been developed.

Cut three slots 65 mm long and 12.5 mm wide in the body of the container. Two small holes 2 mm in diameter are made in the centers of the container bottom and the snap-on lid so that the probe can slide along the anchor line. The retrieval line is knotted through another hole in the lid.

Three packages of *ChexSorb II* resin are fitted inside the probe. The probe is 120 mm high, 95 mm diameter, and has a mass of 120 g.

A weighted anchor line is lowered until the weight rests on the bottom. A probe is threaded onto the anchor line, and a buoy is attached to the top.

The free end of the retrieval line is tied to the buoy. To retrieve the probe, untie the retrieval line and raise the sample to the surface.

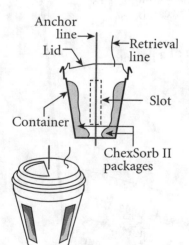

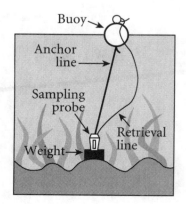

Electricity and Magnetism

SAFETY

Never close a circuit until it has been approved by your teacher. Never rewire or adjust any element of a closed circuit.

Never work with electricity near water; be certain that the floor and all work surfaces are dry.

Do not attempt this exercise with any batteries, electrical devices, or magnets other than those provided by your teacher for this purpose.

OBJECTIVES

* Observe the effects of a current through a wire.
* Discover how the core of an electromagnet affects the magnet's strength.
* Construct a simple speaker.

Exploring magnetic fields around wires

Procedure

1. Leaving a 20 cm tail, as shown below, wind 1.5 m of insulated wire around a cylindrical pen to create 40 tight coils. The coils should touch each other but should not overlap. Leave another 20 cm tail on the other end of the coils, as shown.

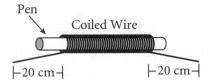

Pen
Coiled Wire
⊢20 cm⊣   ⊢20 cm⊣

2. Keeping the wire coiled, carefully remove the wire from the pen.

3. Using leads with alligator clips, connect the coil to the posts of the battery holder as shown. Carefully insert the batteries in the battery holder.

4. Place the compass directly under the wire leading from the battery to the coil. Observe the position of the compass needle, and record your observations in your lab notebook.

5. Move the compass to a position directly above the wire leading from the battery to the coil. Observe the position of the compass needle, and record your observations in your lab notebook.

MATERIALS

* ✔ headphone plug
* ✔ 2 D-cell batteries
* ✔ brass screw or bolt
* ✔ steel screw or bolt
* ✔ brass nut
* ✔ steel nut
* ✔ battery holders for 2 D-cell batteries
* ✔ ceramic disk magnets
* ✔ cylindrical plastic pen
* ✔ electrical tape
* ✔ film canister with a hole in the bottom
* ✔ magnet wire
* ✔ insulated wire
* ✔ insulated connecting wires with alligator clips
* ✔ masking tape
* ✔ meterstick
* ✔ paper clips
* ✔ portable battery-powered radio
* ✔ metric ruler
* ✔ small compass
* ✔ wire cutters

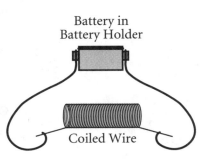

Battery in
Battery Holder

Coiled Wire

6. Place the coil horizontally on the desk. Touch the compass to one opening of the coil. Observe the position of the compass needle, and record your observations in your lab notebook.

7. Touch the compass to the other opening of the coil. Observe the position of the compass needle, and record your observations in your lab notebook.

8. Move the compass from one end of the coil to the other end while keeping it in contact with the side of the coil. Observe the position of the compass needle, and record your observations in your lab notebook.

9. Place 5 or 6 paper clips on the table near the coil. Without moving the battery, carefully move the coil close to the paper clips. Observe the paper clips, and record your observations in your lab notebook.

10. Remove the batteries from the battery holder. Disconnect the coil from the posts of the battery holder.

Analysis

A. Describe how a compass needle responds to a current-carrying wire when the compass needle is below the wire, when the compass needle is above the wire, and when the compass moves from above the wire to below the wire.

Electromagnet cores

Procedure

Nut
Coiled Wire on Bolt
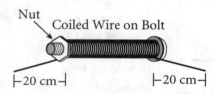
⊢20 cm⊣   ⊢20 cm⊣

11. Insert a steel bolt through the wire coil. Insert the tip of the bolt protruding from the coil into a nut. Screw the nut clockwise, securing the coil on the bolt as shown.

12. Using leads with alligator clips, connect the coil to the posts of the battery holder. Carefully insert the batteries in the battery holder.

13. Touch the compass to one end of the bolt. Move the compass from one end of the bolt to the other end while keeping it in contact with one side of the coil. Observe the position of the compass needle and record your observations in your lab notebook.

14. Place five or six paper clips on the table near the coil. Without moving the battery, carefully move the coil close to the paper clips. Observe the paper clips, and record your observations in your lab notebook.

15. Remove the batteries from the battery holder. Disconnect the coil from the posts of the battery holder.

16. Replace the steel bolt with a brass bolt and repeat steps 11–15.

17. Remove the batteries from the battery holder. Disconnect the coil from the posts of the battery holder.

Analysis

B. Did the wire coil with the bolt have the same effect as the wire coil alone? Explain.

C. What happened as the bolt moved toward the paper clips?

D. Did the coil with the steel bolt pick up the paper clips as effectively as the coil with the brass bolt did?

Constructing a simple speaker

Procedure

18. Leaving a 40 cm tail, wind 2.0 m of magnet wire around a film canister with a hole in its base. The coils should touch each other but should not overlap. After five coils, tape the wire so that it does not fall off the canister. Wrap until you are 0.5 cm from the base of the canister. Carefully tape the coils to the side of the canister so that they do not unravel. Leave another 40 cm tail on the other end of the coils.

19. Place the uncoiled tail of the wire coil on a piece of cardboard. Using the wire cutters, carefully remove the enamel coating on the last 3.0 cm of each end of the wire.

20. Unscrew the casing from the phone plug. Thread both ends of the wire through the hole in the casing. Move this casing 25 cm up the wires.

21. Connect one wire to each of the metal posts of the phone plug. Make sure that opposite wires and posts do not touch. Wrap tape around one of the wires so that it does not touch the other wire.

22. Move the casing back down the wires. Screw the casing onto the metal part of the phone plug.

23. Move to a quiet area. Stack the ceramic magnets flat on the table. Place the film canister over the magnets so that it rests on the table.

24. Tune the portable radio to a station, and decrease the volume to its minimum. Insert the phone plug into the headphones slot on the portable radio.

25. Position your ear on the hole at the film canister base. Slowly increase the volume setting on the radio. You should be able to hear the radio. If not, reopen the phone plug and check the connections.

Analysis

E. What powers the electromagnet in the speaker?

F. Different parts of a speaker must pull and push on each other to produce sound waves that travel to your ear. Describe how different parts of this speaker produce sound.

G. Describe some ways to get the speaker to produce a louder sound.

Building a Circuit Breaker

Miriam Parsa
20350 Via Subida
San Diego, California 92117

May 4, 2000

Dr. Shana Gillis
1% Inspiration Laboratories
14557 West Post Road
Tempe, Arizona 85289

Dear Dr. Gillis:

While looking through a trunk filled with my deceased mother's belongings, I came across a lab notebook with instructions for a circuit breaker. I am excited about this find because the breaker she designed appears very cheap to make; it can be made from materials commonly found around the home. It also should sell well because a circuit breaker usually works much better than a fuse.

I am enclosing all the instructions that my mother left for the working part of the circuit breaker. The problem is that her instructions are incomplete, and I have not been able to build this device or test it in a working circuit. I need your help in completing the breaker and in figuring out how to install it in a circuit. I do know that a moving part is missing from the design. I also know that the moving part must somehow connect and then disconnect the current-carrying wires. Can you and your staff help me? Please let me know as soon as possible. Thank you so much for your time.

Sincerely,

Miriam Parsa

Miriam Parsa

The page from the notebook is on page 100.

1% INSPIRATION LABORATORIES

MEMORANDUM

Date: June 21, 2000
To: Research and Development
From: Shana Gillis

The notes provide us with a good start on the solenoid. First, write up a plan for building the circuit breaker and fitting it in the circuit.

After I approve your plan, go into the lab and follow these steps:

1. Make the part of the solenoid described in the notes. Follow the directions carefully.

2. Decide what material to use for a plunger, and determine how the plunger should be shaped.

3. Get the solenoid-and-plunger device to work when power is applied directly to it.

4. Mount the solenoid and plunger onto a piece of cardboard, and wire the device in a circuit consisting of a bulb and a battery pack. Show that the circuit works well under normal conditions but that when a short circuit occurs and there is too much current, the solenoid shuts off the current.

I would like a drawing of the complete circuit and a short explanation of what you expect to happen. I have included a list of equipment that we have available.

You will prepare your report in the form of a patent application.

14557 West Post Road • Tempe, Arizona 852⬚

See next page for safety requirements, materials list, and more hints.

MATERIALS

ITEM	QTY.
✔ large metal paper clips	1 box
✔ magnet wire	1 roll
✔ plastic drinking straw	1
✔ cardboard	1
✔ battery pack for 2 D-cells	1
✔ D-cell battery	2
✔ lamp board with 5 miniature sockets	1
✔ miniature bulb (3 V)	5
✔ craft knife	1
✔ electrical tape	1 roll
✔ scissors	1
✔ aluminum foil	1
✔ modeling clay	
✔ connecting leads with alligator clips	3
✔ switch	1
✔ bare copper wire	70 cm

SAFETY

Wire coils may heat rapidly during this experiment. If heating occurs, open the switch immediately and handle the equipment with a hot mitt. Allow all equipment to cool before storing it.

Never close a circuit until it has been approved by your teacher. Never rewire or adjust any element of a closed circuit. Never work with electricity near water; be sure the floor and all work surfaces are dry.

Do not attempt this exercise with any batteries or electrical devices other than those provided by your teacher for this purpose.

Date: October 27, 1957

A solenoid consists of many coils of wire neatly wrapped around a hollow cylinder. A thin piece of magnetic metal (such as iron, steel, or nickel) is then placed partly in the cylinder. This metal acts as a plunger. When current is put through the coils of wire, the magnetic field created pulls the plunger completely into the cylinder. I think I can use this device to make a circuit breaker.

Measure and mark 6 cm from the end of a plastic drinking straw. Measure a 10 cm tail at the end of a magnet wire, and bend the wire 90°. Tape the tail to the long part of the straw so that the bend is at the 6 cm mark. Wind tight, even coils of wire from the bend to the free end of the straw, making sure the coils touch each other. Count the coils as you wrap. Tape the coils down at regular intervals. When you reach the end, place tape over the first layer of coils, and begin wrapping in the opposite direction. Continue wrapping back and forth down the length of the straw until you have 200 coils (this should make at least four layers of wire on top of the straw). Leave a 5 cm tail of wire when you finish. Now take a